Darwin
on
Trial

Darwin on Trial

Phillip E. Johnson

INTERVARSITY PRESS
DOWNERS GROVE, ILLINOIS 60515

InterVarsity Press
P.O. Box 1400, Downers Grove, Illinois 60515

InterVarsity Press is the book-publishing division of InterVarsity Christian
Fellowship, a student movement active on campus at hundreds of universities,
colleges and schools of nursing in the United States of America, and a member
movement of the International Fellowship of Evangelical Students. For
information about local and regional activities, write Public Relations Dept.,
InterVarsity Christian Fellowship, 6400 Schroeder Rd., P.O. Box 7895, Madison,
WI 53707-7895.

ISBN 0-8308-1758-1
Printed in the United States of America

Library of Congress Cataloging-in-Publication Data
Johnson, Phillip E., 1940–
    Darwin on trial / Phillip E. Johnson.
      p.   cm.
    Includes bibliographical references and index.
    ISBN 0-89526-535-4 (alk. paper)
    1. Evolution.   I. Title.
  QH366.2.J65   1991
  575—dc20   90-26218
                                   CIP

15 14 13 12 11 10 9 8 7 6 5 4 3
01 00 99 98 97 96 95 94 93 92 91

To those (especially Kathie) who listened and read,
and did their best to help me stay
on the straight and narrow path; and

To those brave souls who asked the hard questions
even when there was never a chance
of getting a straight answer; and

To those in science who want to allow
the questions to be asked.

Contents

Darwin
on
Trial

Chapter One

The Legal Setting

IN 1981 THE STATE legislature of Louisiana passed a law requiring that if "evolution-science" is taught in the public schools, the schools must also provide balanced treatment for something called "creation-science." The statute was a direct challenge to the scientific orthodoxy of today, which is that all living things evolved by a gradual, natural process—from nonliving matter to simple microorganisms, leading eventually to man. Evolution is taught in the public schools (and presented in the media) not as a theory but as a fact, the "fact of evolution." There are nonetheless many dissidents, some with advanced scientific degrees, who deny that evolution is a fact and who insist that an intelligent Creator caused all living things to come into being in furtherance of a purpose.

The conflict requires careful explanation, because the terms are confusing. The concept of creation in itself does not imply opposition to evolution, if evolution means only a gradual process by which one kind of living creature changes into something different. A

3

Creator might well have employed such a gradual process as a means of creation. "Evolution" contradicts "creation" only when it is explicitly or tacitly defined as *fully naturalistic evolution*—meaning evolution that is not directed by any purposeful intelligence.

Similarly, "creation" contradicts evolution only when it means *sudden creation*, rather than creation by progressive development. For example, the term "creation-science," as used in the Louisiana law, is commonly understood to refer to a movement of Christian fundamentalists based upon an extremely literal interpretation of the Bible. Creation-scientists do not merely insist that life was *created*; they insist that the job was completed in six days no more than ten thousand years ago, and that all evolution since that time has involved trivial modifications rather than basic changes. Because creation-science has been the subject of so much controversy and media attention, many people assume that anyone who advocates "creation" endorses the "young earth" position and attributes the existence of fossils to Noah's flood. Clearing up that confusion is one of the purposes of this book.[1]

The Louisiana statute and comparable laws in other states grew out of the long-standing efforts of Christian fundamentalists to reassert the scientific vitality of the Biblical narrative of creation against its Darwinist rival. The great landmark in this Bible-science conflict was the famous *Scopes* case, the "monkey trial" of the 1920s, which most Americans know in the legendary version portrayed in the play and movie *Inherit the Wind*. The legend tells of religious fanatics who invade a school classroom to persecute an inoffensive science teacher, and of a heroic defense lawyer who symbolizes reason itself in its endless battle against superstition.

As with many legendary incidents the historical record is more complex. The Tennessee legislature had passed as a symbolic measure a statute prohibiting the teaching of evolution, which the

---

[1] Clearing up confusion requires a careful and consistent use of terms. In this book, "creation-science" refers to young-earth, six-day special creation. "Creationism" means belief in creation in a more general sense. Persons who believe that the earth is billions of years old, and that simple forms of life evolved gradually to become more complex forms including humans, are "creationists" if they believe that a supernatural Creator not only initiated this process but in some meaningful sense *controls* it in furtherance of a purpose. As we shall see, "evolution" (in contemporary scientific usage) excludes not just creation-science but creationism in the broad sense. By "Darwinism" I mean fully naturalistic evolution, involving chance mechanisms guided by natural selection.

governor signed only with the explicit understanding that the ban would not be enforced. Opponents of the law (and some people who just wanted to put Dayton, Tennessee, on the map) engineered a test case. A former substitute teacher named Scopes, who wasn't sure whether he had ever actually *taught* evolution, volunteered to be the defendant.

The case became a media circus because of the colorful attorneys involved. William Jennings Bryan, three-time Democratic presidential candidate and secretary of state under President Woodrow Wilson, led the prosecution. Bryan was a Bible believer but not an uncompromising literalist, in that he thought that the "days" of Genesis referred not to 24-hour periods but to historical ages of indefinite duration. He opposed Darwinism largely because he thought that its acceptance had encouraged the ethic of ruthless competition that underlay such evils as German militarism and robber baron capitalism.

The Scopes defense team was led by the famous criminal lawyer and agnostic lecturer Clarence Darrow. Darrow maneuvered Bryan into taking the stand as an expert witness on the Bible and humiliated him in a devastating cross-examination. Having achieved his main purpose, Darrow admitted that his client had violated the statute and invited the jury to convict. The trial thus ended in a conviction and a nominal fine of $100. On appeal, the Tennessee supreme court threw out the fine on a technicality but held the statute constitutional. From a legal standpoint the outcome was inconclusive, but as presented to the world by the sarcastic journalist H. L. Mencken, and later by Broadway and Hollywood, the "monkey trial" was a public relations triumph for Darwinism.

The scientific establishment was not exactly covering itself with glory at the time, however. Although he did not appear at the trial, the principal spokesman for evolution during the 1920s was Henry Fairfield Osborn, Director of the American Museum of Natural History. Osborn relied heavily upon the notorious Piltdown Man fossil, now known to be a fraud, and he was delighted to confirm the discovery of a supposedly pre-human fossil tooth by the paleontologist Harold Cooke in Bryan's home state of Nebraska. Thereafter Osborn prominently featured "Nebraska Man" (scientific designation: *Hesperopithecus haroldcookii*) in his antifundamentalist newspaper articles and radio broadcasts, until the tooth was discovered to be from a peccary, a kind of pig. If Osborn had been cross-examined by a lawyer as clever as Clarence Darrow, and satirized by a columnist as ruthless as H. L. Mencken, he would have looked as silly as Bryan.

The anti-evolution statutes of the 1920s were not enforced, but textbook publishers tended to say as little as possible about evolution to avoid controversy. The Supreme Court eventually held the statutes unconstitutional in 1968, but by then the fundamentalists had changed their objective. Creation research institutes were founded, and books began to appear which attacked the orthodox interpretation of the scientific evidence and argued that the geological and fossil record could be harmonized with the Biblical account. None of this literature was taken seriously by the scientific establishment or the mass media, but the creation-scientists themselves became increasingly confident that they had a scientific case to make.

They also began to see that it was possible to turn the principles of liberal constitutional law to their advantage by claiming a right to debate evolutionists on equal terms in school science classes. Their goal was no longer to suppress the teaching of evolution, but to get a fair hearing for their own viewpoint. If there is a case to be made for both sides of a scientific controversy, why should public school students, for example, hear only one side? Creation-scientists emphasized that they wanted to present only the *scientific* arguments in the schools; the Bible itself was not to be taught.

Of course mainstream science does not agree that there are two sides to the controversy, and regards creation-science as a fraud. Equal time for creation-science in biology classes, the Darwinists like to say, is like equal time for the theory that it is the stork that brings babies. But the consensus view of the scientific establishment is not enshrined in the Constitution. Lawmakers are entitled to act on different assumptions, at least to the extent that the courts will let them.

Louisiana's statute never went into effect because a federal judge promptly held it unconstitutional as an "establishment of religion." In 1987 the Supreme Court of the United States affirmed this decision by a seven to two majority. The Louisiana law was unconstitutional, said the majority opinion by Justice William Brennan, because its purpose "was clearly to advance the religious viewpoint that a supernatural being created humankind." Not so, said the dissenting opinion by Justice Antonin Scalia, because "The people of Louisiana, including those who are Christian fundamentalists, are quite entitled, as a secular matter, to have whatever scientific evidence there may be against evolution presented in their schools, just as Mr. Scopes was entitled to present whatever scientific evidence there was for it."

Both Justice Brennan and Justice Scalia were in a sense right. The Constitution excludes religious advocacy from public school classrooms, and to say that a supernatural being created mankind is certainly to advocate a religious position. On the other hand, the Louisiana legislature had acted on the premise that legitimate scientific objections to "evolution" were being suppressed. Some might doubt that such objections exist, but the Supreme Court could not overrule the legislature's judgment on a disputed scientific question, especially considering that the state had been given no opportunity to show what balanced treatment would mean in practice. In addition, the creation-scientists were arguing that the teaching of evolution itself had a religious objective, namely to *discredit* the idea that a supernatural being created mankind. Taking all this into account, Justice Scalia thought that the Constitution permitted the legislature to give people offended by the allegedly dogmatic teaching of evolution a fair opportunity to reply.

As a legal scholar, one point that attracted my attention in the Supreme Court case was the way terms like "science" and "religion" are used to imply conclusions that judges and educators might be unwilling to state explicitly. If we say that naturalistic evolution is *science*, and supernatural creation is *religion*, the effect is not very different from saying that the former is true and the latter is fantasy. When the doctrines of science are taught as fact, then whatever those doctrines exclude cannot be true. By the use of labels, objections to naturalistic evolution can be dismissed without a fair hearing.

My suspicions were confirmed by the "friend of the court" argument submitted by the influential National Academy of Sciences, representing the nation's most prestigious scientists. Creation-science is not science, said the Academy in its argument to the Supreme Court, because

it fails to display the most basic characteristic of science: reliance upon naturalistic explanations. Instead, proponents of "creation-science" hold that the creation of the universe, the earth, living things, and man was accomplished through supernatural means inaccessible to human understanding.

Because creationists cannot perform scientific research to establish the reality of supernatural creation—that being by definition impossible—the Academy described their efforts as aimed primarily at discrediting evolutionary theory.

"Creation-science" is thus manifestly a device designed to dilute the persuasiveness of the theory of evolution. The dualistic mode of analysis and the negative argumentation employed to accomplish this dilution is, moreover, antithetical to the scientific method.

The Academy thus defined "science" in such a way that advocates of supernatural creation may neither argue for their own position nor dispute the claims of the scientific establishment. That may be one way to win an argument, but it is not satisfying to anyone who thinks it possible that God really did have something to do with creating mankind, or that some of the claims that scientists make under the heading of "evolution" may be false.

I approach the creation-evolution dispute not as a scientist but as a professor of law, which means among other things that I know something about the ways that words are used in arguments. What first drew my attention to the question was the way the rules of argument seemed to be structured to make it impossible to question whether what we are being told about evolution is really true. For example, the Academy's rule against negative argument automatically eliminates the possibility that science has not discovered how complex organisms could have developed. However wrong the current answer may be, it stands until a better answer arrives. It is as if a criminal defendant were not allowed to present an alibi unless he could also show who did commit the crime.

A second point that caught my attention was that the very persons who insist upon keeping religion and science separate are eager to use their science as a basis for pronouncements about religion. The literature of Darwinism is full of anti-theistic conclusions, such as that the universe was not designed and has no purpose, and that we humans are the product of blind natural processes that care nothing about us. What is more, these statements are not presented as personal opinions but as the logical implications of evolutionary science.

Another factor that makes evolutionary science seem a lot like religion is the evident zeal of Darwinists to evangelize the world, by insisting than even non-scientists accept the truth of their theory as a matter of moral obligation. Richard Dawkins, an Oxford Zoologist who is one of the most influential figures in evolutionary science, is unabashedly explicit about the religious side of Darwinism. His 1986 book *The Blind Watchmaker* is at one level about biology, but at a more fundamental level it is a sustained argument for atheism. According to Dawkins, "Darwin made it possible to be an intellectually fulfilled atheist."

When he contemplates the perfidy of those who refuse to believe, Dawkins can scarcely restrain his fury. "It is absolutely safe to say that, if you meet somebody who claims not to believe in evolution, that person is ignorant, stupid or insane (or wicked, but I'd rather not consider that)." Dawkins went on to explain, by the way, that what he dislikes particularly about creationists is that they are intolerant.

We must therefore believe in evolution or go to the madhouse, but what precisely is it that we are required to believe? "Evolution" can mean anything from the uncontroversial statement that bacteria "evolve" resistance to antibiotics to the grand metaphysical claim that the universe and mankind "evolved" entirely by purposeless, mechanical forces. A word that elastic is likely to mislead, by implying that we know as much about the grand claim as we do about the small one.

That very point was the theme of a remarkable lecture given by Colin Patterson at the American Museum of Natural History in 1981. Patterson is a senior paleontologist at the British Natural History Museum and the author of that museum's general text on evolution. His lecture compared creationism (not creation-science) with evolution, and characterized both as scientifically vacuous concepts which are held primarily on the basis of faith. Many of the specific points in the lecture are technical, but two are of particular importance for this introductory chapter. First, Patterson asked his audience of experts a question which reflected his own doubts about much of what has been thought to be secure knowledge about evolution:

Can you tell me anything you know about evolution, any one thing . . . that is true? I tried that question on the geology staff at the Field Museum of Natural History and the only answer I got was silence. I tried it on the members of the Evolutionary Morphology seminar in the University of Chicago, a very prestigious body of evolutionists, and all I got there was silence for a long time and eventually one person said "I do know one thing—it ought not to be taught in high school."

Patterson suggested that both evolution and creation are forms of pseudo-knowledge, concepts which seem to imply information but do not. One point of comparison was particularly striking. A common objection to creationism in pre-Darwinian times was that no one could say anything about the mechanism of creation. Creationists simply pointed to the "fact" of creation and conceded ignorance of the means. But now, according to Patterson, Darwin's theory of natural selection is under fire and scientists are no longer sure of its general validity. Evolutionists increasingly talk like creationists in that they point to a fact but cannot provide an explanation of the means.

Patterson was being deliberately provocative, and I do not mean to imply that his skeptical views are widely supported in the scientific community. On the contrary, Patterson came under heavy fire from Darwinists after somebody circulated a bootleg transcript of the lecture, and he eventually disavowed the whole business. Whether or not he meant to speak for public attribution, however, he was making an important point. We can point to a mystery and call it "evolution," but this is only a label. The important question is not whether scientists have agreed on a label, but how much they know about how complex living beings like ourselves came into existence.

Irving Kristol is a prominent social theorist with a talent for recognizing ideological obfuscation, and he applied that talent to Darwinism in an essay in *The New York Times*. Kristol observed that Darwinian theory, which explains complex life as the product of small genetic mutations and "survival of the fittest," is known to be valid only for variations within the biological species. That Darwinian evolution can gradually transform one kind of creature into another is merely a biological hypothesis, not a fact. He noted that science abounds with rival opinions about the origin of life and that some scientists have questioned whether the word "evolution" carries much meaning. Kristol conceded that creation-science is a matter of faith and not science, and should not be taught in the schools, but he thought that its defenders still had a point:

> It is reasonable to suppose that if evolution were taught more cautiously, as a conglomerate idea consisting of conflicting hypotheses rather than as an unchallengeable certainty, it would be far less controversial. As things now stand, the religious fundamentalists are not far off the mark when they assert that evolution, as generally taught, has an unwarranted anti-religious edge to it.

One famous evolutionist who might have been expected to be sympathetic to Kristol's point would be Harvard Professor Stephen Jay Gould. In 1980 Gould published a paper in a scientific journal predicting the emergence of "a new and general theory of evolution" to replace the neo-Darwinian synthesis. Gould wrote that, although he had been "beguiled" by the unifying power of the Darwinist synthesis when he studied it as a graduate student in the 1960s, the weight of the evidence had driven him to the reluctant conclusion that the synthesis, "as a general proposition, is effectively dead, despite its persistence as textbook orthodoxy." The dogmatic teaching of that dead textbook orthodoxy was precisely what Kristol was criticizing.

Gould nonetheless wrote a reply to Kristol that put this outsider firmly in his place. Gould denied that textbook bias was more prevalent in evolution than in other fields of science, denied that evolutionary science is anti-religious, and insisted that "Darwinian selection . . . will remain a central focus of more inclusive evolutionary theories." His main point was that Kristol had ignored a "central distinction between secure fact and healthy debate about theory." Biologists do teach evolutionary *theory* as a conglomerate idea consisting of conflicting hypotheses, Gould wrote, but evolution is also a *fact* of nature, as well established as the fact that the earth revolves around the sun.[2]

As an outside observer who enjoys following the literature of evolution and its conflicts, I have become accustomed to seeing this sort of evasive response to criticism. When outsiders question

---

[2] Gould's arguments for the "fact of evolution" are the subject of Chapters Five and Six of this book.

whether the theory of evolution is as secure as we have been led to believe, we are firmly told that such questions are out of order. The arguments among the experts are said to be about matters of detail, such as the precise timescale and mechanism of evolutionary transformations. These disagreements are signs not of crisis but of healthy creative ferment within the field, and in any case there is no room for doubt whatever about something called the "fact" of evolution.

But consider Colin Patterson's point that a fact of evolution is vacuous unless it comes with a supporting theory. Absent an explanation of how fundamental transformations can occur, the bare statement that "humans evolved from fish" is not impressive. What makes the fish story impressive, and credible, is that scientists think they know how a fish can be changed into a human without miraculous intervention.

Charles Darwin made evolution a scientific concept by showing, or claiming to have shown, that major transformations could occur in very small steps by purely natural means, so that time, chance, and differential survival could take the place of a miracle. If Darwin's scenario of gradual adaptive change is wrong, then "evolution" may be no more than a label we attach to the observation that men and fish have certain common features, such as the vertebrate body plan.

Disagreements about the mechanism of evolution are therefore of fundamental importance to those of us who want to know whether the scientists really know as much as they have been claiming to know. An adequate theory of how evolution works is particularly indispensable when evolution is deemed to imply, as countless Darwinists have insisted, that purposeless material mechanisms are responsible for our existence. "Evolution" in the sense in which these scientists use the term *is* a mechanistic process, and so the content of any "fact" that is left when the mechanism is subtracted is thoroughly obscure.

In the chapters to follow I will look at the evidence to see whether a mechanism is known that can accomplish the large-scale changes which the theory of evolution supposes to have occurred, such as the change from single-celled bacteria to complex plants and animals, from fish to mammals, and from apes to men. If the neo-Darwinist mechanism will not do the job, and if instead of an established replacement we have only what Gould and Kristol agreed to call "a conglomerate idea consisting of conflicting hypotheses," then we may conclude that the scientists do not in fact know how large-scale evolution could have occurred. We will then have to consider whether a "fact of evolution" can be separated from Darwin's theory. Our investigation will require us to explore the new evidence revealed by molecular studies, the state of research into the origin of life, and the rules of scientific inquiry.

Before undertaking this task I should say something about my qualifications and purpose. I am not a scientist but an academic lawyer by profession, with a specialty in analyzing the logic of arguments and identifying the assumptions that lie behind those arguments. This background is more appropriate than one might think, because what people believe about evolution and Darwinism depends very heavily on the kind of logic they employ and the kind of assumptions they make.[3] Being a scientist is not necessarily an advantage when dealing with a very broad topic like evolution, which cuts across many scientific disciplines and also involves issues of philosophy. Practicing scientists are of necessity highly specialized, and a scientist outside his field of expertise is just another layman.

Access to the relevant scientific information presents no great difficulty. Charles Darwin and T. H. Huxley wrote for the general reader, and the same is true of the giants of the neo-Darwinist synthesis such as Theodosius Dobzhansky, George Gaylord Simpson, and Julian Huxley. Current authors who address the general public and who are eminent among scientists include Stephen Jay Gould, Richard Dawkins, Douglas Futuyma, and a host of other experts who are named in the research notes to each chapter.

Most of the professional scientific literature is available in the premier scientific journals *Nature* and *Science*, the most prestigious scientific organs in Britain and America respectively, and at a somewhat more popular level in the British *New Scientist* and the *Scientific American*. Philosophers and historians have also produced well-informed books. In short the available literature is voluminous, and the leading scientific figures have always assumed that nonscientist

---

[3] When the National Academy of Sciences appointed a special committee to prepare its official booklet titled *Science and Creationism*, four of the eleven members were lawyers.

readers can understand the essential evidence. But evidence never speaks for itself; it has meaning only in the context of rules of reasoning which determine what may be considered and what counts as evidence. Those rules of reasoning are what I particularly want to examine.

The last subject I should address before beginning is my personal religious outlook, because readers are bound to wonder and because I do not exempt myself from the general rule that bias must be acknowledged and examined. I am a philosophical theist and a Christian. I believe that a God exists who could create out of nothing if He wanted to do so, but who might have chosen to work through a natural evolutionary process instead. I am not a defender of creation-science, and in fact I am not concerned in this book with addressing any conflicts between the Biblical accounts and the scientific evidence.

My purpose is to examine the scientific evidence on its own terms, being careful to distinguish the evidence itself from any religious or philosophical bias that might distort our interpretation of that evidence. I assume that the creation-scientists are biased by their precommitment to Biblical fundamentalism, and I will have very little to say about their position. The question I want to investigate is whether Darwinism is based upon a fair assessment of the scientific evidence, or whether it is another kind of fundamentalism.

Do we really know for certain that there exists some natural process by which human beings and all other living beings could have evolved from microbial ancestors, and eventually from non-living matter? When the National Academy of Sciences tells us that reliance upon naturalistic explanations is the most basic characteristic of science, is it implying that scientists somehow know that a Creator played no part in the creation of the world and its forms of life? Can something be non-science but true, or does non-science mean non*sense*? Given the emphatic endorsement of naturalistic evolution by the scientific community, can outsiders even contemplate the possibility that this officially established doctrine might be false? Well, come along and let us see.

Chapter Two

Natural Selection

THE STORY OF Charles Darwin has been told many times, and no wonder. The relationship with the lawyer-geologist Charles Lyell, the long voyage in the Beagle with the temperamental Captain Fitzroy, the observations and adventures in South America and the Galapagos Islands, the long years of preparation and delay, the eventual rushed publication of *The Origin of Species* when Alfred Russell Wallace appeared about to publish a similar theory, the controversies and the smashing triumph—all these make a great saga which is always worth another retelling. My subject is not history but the logic of current controversy, however, and so my interest must be in Darwinism and not Darwin. I am also uninterested in the differences between the theory as Darwin originally proposed it and as it is understood by neo-Darwinists today, who have the advantage of the greater understanding of genetics that science has achieved since Darwin's time. My purpose is to explain what concepts the contemporary theory employs, what significant statements about the natural world it makes, and what points of legitimate controversy there may be.

Darwin's classic book argued three important related propositions. The first was that "the species are not immutable." By this he meant that new species have appeared during the long course of the earth's history by a natural process he called "descent with modification." The second proposition was that this evolutionary process can be extended to account for all or nearly all the diversity of life, because all living things descended from a very small number of common ancestors, perhaps a single microscopic ancestor. The third proposition, and the one most distinctive to Darwinism, was that this vast process was guided by natural selection or "survival of the fittest," a guiding force so effective that it could accomplish prodigies of biological craftsmanship that people in previous times had thought to require the guiding hand of a creator.[1] The evidence for this third proposition is the subject of this chapter.

The question is *not* whether natural selection occurs. Of course it does, and it has an effect in maintaining the genetic fitness of a population. Infants with severe birth defects do not survive to maturity without expensive medical care, and creatures which do not survive to reproduce do not leave descendants. These effects are unquestioned, but Darwinism asserts a great deal more than merely that species avoid genetic deterioration due to natural attrition among the genetically unfit. Darwinists claim that this same force of attrition has a building effect so powerful that it can begin with a bacterial cell and gradually craft its descendants over billions of years to produce such wonders as trees, flowers, ants, birds, and humans. How do we know that all this is possible?

Darwinian evolution postulates two elements. The first is what Darwin called "variation," and what scientists today call *mutation*.[2]

---

[1] Darwin did not insist that all evolution was by natural selection, nor do his successors. He wrote at the end of the introduction to the first (1859) edition of *The Origin of Species* that "I am convinced that natural selection has been the main but not the exclusive means of modification" and later complained of the "steady misrepresentation" that had ignored this qualification. On the other hand, Darwin was vague about the importance of the alternatives, one of which was "variations which seem to us in our ignorance to arise spontaneously." Contemporary neo-Darwinists also practice a tactically advantageous flexibility concerning the frequency and importance of non-selective evolution. Stephen Jay Gould wrote that this imprecision "imposes a great frustration upon anyone who would characterize the modern synthesis in order to criticize it," and I am sure that every critic shares the frustration. Readers should therefore beware of taking at face value claims by neo-Darwinist authorities that some critic has misunderstood or mischaracterized their theory.

[2] "Mutation" as used here is a simple label for the set of mechanisms which provide the genetic variation upon which natural selection can go to work. The set includes point mutations, chromosomal doubling, gene duplication, and recombination. The essential point

Mutations are randomly occurring genetic changes which are nearly always harmful when they produce effects in the organism large enough to be visible, but which may occasionally slightly improve the organism's ability to survive and reproduce. Organisms generally produce more offspring than can survive to maturity, and offspring that possess an advantage of this kind can be expected to produce more descendants themselves, other things being equal, than less advantaged members of the species. As the process of differential survival continues, the trait eventually spreads throughout the species, and it may become the basis for further cumulative improvements in succeeding generations. Given enough time, and sufficient mutations of the right sort, enormously complex organs and patterns of adaptive behavior can eventually be produced in tiny cumulative steps, without the assistance of any pre-existing intelligence.

That is, all this can happen if the theory is true. Darwin could not point to impressive examples of natural selection in action, and so he had to rely heavily on an argument by analogy. In the words of Douglas Futuyma:

> When Darwin wrote *The Origin of Species*, he could offer no good cases of natural selection because no one had looked for them. He drew instead an analogy with the artificial selection that animal and plant breeders use to improve domesticated varieties of animals and plants. By breeding only from the woolliest sheep, the most fertile chickens, and so on, breeders have been spectacularly successful in altering almost every imaginable characteristic of our domesticated animals and plants to the point where most of them differ from their wild ancestors far more than related species differ from them.

The analogy to artificial selection is misleading. Plant and animal breeders employ intelligence and specialized knowledge to select breeding stock and to protect their charges from natural dangers. The point of Darwin's theory, however, was to establish that purposeless natural processes can substitute for intelligent design.

---

is that the variations are supposed to be *random*. Creative evolution would be much easier to envisage if some guiding force caused the right mutations to arrive on schedule. Orthodox genetic theory insists that no such guiding principle for mutation exists, so creatures have to make do with whatever blind nature happens to provide.

That he made that point by citing the accomplishments of intelligent designers proves only that the receptive audience for his theory was highly uncritical.

Artificial selection is not basically the same sort of thing as natural selection, but rather is something fundamentally different. Human breeders produce variations among sheep or pigeons for purposes absent in nature, including sheer delight in seeing how much variation can be achieved. If the breeders were interested only in having animals capable of surviving in the wild, the extremes of variation would not exist. When domesticated animals return to the wild state, the most highly specialized breeds quickly perish and the survivors revert to the original wild type. Natural selection is a conservative force that prevents the appearance of the extremes of variation that human breeders like to encourage.

What artificial selection actually shows is that there are definite limits to the amount of variation that even the most highly skilled breeders can achieve. Breeding of domestic animals has produced no new species, in the commonly accepted sense of new breeding communities that are infertile when crossed with the parent group. For example, all dogs form a single species because they are chemically capable of interbreeding, although inequality of size in some cases makes natural copulation impracticable. The eminent French zoologist Pierre Grassé concluded that the results of artificial selection provide powerful testimony against Darwin's theory:

> In spite of the intense pressure generated by artificial selection (eliminating any parent not answering the criteria of choice) over whole millennia, no new species are born. A comparative study of sera, hemoglobins, blood proteins, interfertility, etc., proves that the strains remain within the same specific definition. This is not a matter of opinion or subjective classification, but a measurable reality. The fact is that selection gives tangible form to and gathers together all the varieties a genome is capable of producing, but does not constitute an innovative evolutionary process.

In other words, the reason that dogs don't become as big as elephants, much less change into elephants, is not that we just haven't been breeding them long enough. Dogs do not have the genetic capacity for that degree of change, and they stop getting bigger when the genetic limit is reached.

Darwinists disagree with that judgment, and they have some points to make. They point with pride to experiments with laboratory fruitflies. These have not produced anything but fruitflies, but they have produced changes in a multitude of characteristics. Plant hybrids have been developed which can breed with each other, but not with the parent species, and which therefore meet the accepted standard for new species. With respect to animals, Darwinists attribute the inability to produce new species to a lack of sufficient time. Humans have been breeding dogs for only a few thousand years, but nature has millions and even hundreds of millions of years at her disposal. In some cases, convincing circumstantial evidence exists of evolution that has produced new species in nature. Familiar examples include the hundreds of fruitfly species in Hawaii and the famous variations among "Darwin's Finches" on the Galapagos Islands.

The time available unquestionably has to be taken into account in evaluating the results of breeding experiments, but it is also possible that the greater time available to nature may be more than counterbalanced by the power of intelligent purpose which is brought to bear in artificial selection. With respect to the famous fruitfly experiments, for example, Grassé noted that "The fruitfly (*drosophila melanogaster*) the favorite pet insect of the geneticists, whose geographical, biotropical, urban, and rural genotypes are now known inside out, seems not to have changed since the remotest times." Nature has had plenty of time, but it just hasn't been doing what the experimenters have been doing.

Lack of time would be a reasonable excuse if there were no other known factor limiting the change that can be produced by selection, but in fact selective change is limited by the inherent variability in the gene pool. After a number of generations the capacity for variation runs out. It might conceivably be renewed by mutation, but whether (and how often) this happens is not known.

Whether selection has ever accomplished speciation (i.e. the production of a new species) is not the point. A biological species is simply a group capable of interbreeding. Success in dividing a fruitfly population into two or more separate populations that cannot interbreed would not constitute evidence that a similar process could in time produce a fruitfly from a bacterium. If breeders one day did succeed in producing a group of dogs that can reproduce with each other but not with other dogs, they would still have made only the tiniest step towards proving Darwinism's important claims.

That the analogy to artificial selection is defective does not necessarily mean that Darwin's theory is wrong, but it does mean that we will have to look for more direct evidence to see if natural selection really does have a creative effect. *Before* looking at what the Darwinists have been able to come up with, however, we need to ask whether evidence is even necessary. Strange as it may seem, there are many statements in the Darwinist literature to the effect that the validity of the theory can be demonstrated simply as a matter of logic.

Natural Selection as a Tautology

Many of the most prominent neo-Darwinists have written at one time or another that natural selection is a tautology, a way of saying the same thing twice. In this formulation the theory predicts that the fittest organisms will produce the most offspring, and it defines the fittest organisms as the ones which produce the most offspring. It is important to document this point, because many Darwinists have convinced themselves that the tautology idea is a misunderstanding introduced into the literature by creationists and other uncomprehending faultfinders. But here are a few examples collected by Norman Macbeth:

J. B. S. Haldane (1935): " . . . the phrase, 'survival of the fittest,' is something of a tautology. So are most mathematical theorems. There is no harm in saying the same truth in two different ways."

Ernst Mayr (1963): " . . . those individuals that have the most offspring are by definition . . . the fittest ones."

George Gaylord Simpson (1964): "Natural selection favors fitness only if you define fitness as leaving more descendants. In fact geneticists do define it that way, which may be confusing to others. To a geneticist fitness has nothing to do with health, strength, good looks, or anything but effectiveness in breeding."

The explanation by Simpson just quoted indicates why it is not easy to formulate the theory of natural selection other than as a tautology. It may seem obvious, for example, that it is advantageous for a wild stallion to be able to run faster, but in the Darwinian sense this will be true only to the extent that a faster stallion sires more offspring. If greater speed leads to more frequent falls, or if faster stallions tend to outdistance the mares and miss opportunities for reproduction, then the improvement may be disadvantageous.

Just about any characteristic can be either advantageous or disadvantageous, depending upon the surrounding environmental conditions. Does it seem that the ability to fly is obviously an advantage? Darwin hypothesized that natural selection might have caused beetles on Madeira to lose the ability to fly, because beetles capable of flight tended to be blown out to sea. The large human brain requires a large skull which causes discomfort and danger to the mother in childbirth. We assume that our brain size is advantageous because civilized humans dominate the planet, but it is far from obvious that the large brain was a net advantage in the circumstances in which it supposedly evolved. Among primates in general, those with the largest brains are not the ones least in danger of extinction.

In all such cases we can presume a characteristic to be advantageous because a species which has it seems to be thriving, but in most cases it is impossible to identify the advantage independently of the outcome. That is why Simpson was so insistent that "advantage" has no inherent meaning other than actual success in reproduction. All we can say is that the individuals which produced the most offspring must have had the qualities required for producing the most offspring.

The famous philosopher of science Karl Popper at one time wrote that Darwinism is not really a scientific theory because natural selection is an all-purpose explanation which can account for anything, and which therefore explains nothing. Popper backed away from this position after he was besieged by indignant Darwinist protests, but he had plenty of justification for taking it. As he wrote in his own defense, "some of the greatest contemporary Darwinists themselves formulate the theory in such a way that it amounts to the tautology that those organisms that leave most offspring leave most offspring," citing Fisher, Haldane, Simpson, "and others." One of the others was C. H. Waddington, whose attempt to make sense of the matter deserves to be preserved for posterity:

Darwin's major contribution was, of course, the suggestion that evolution can be explained by the natural selection of random variations. Natural selection, which was at first considered as though it were a hypothesis that was in need of experimental or observational confirmation, turns out on closer inspection to be a tautology, a statement of an inevitable but previously unrecognized relation. It states that the fittest individuals in a population (defined as those which leave most offspring) will leave most offspring. This fact in no way reduces the magnitude of Darwin's achievement; only after it was clearly formulated, could biologists realize the enormous power of the principle as a weapon of explanation.

That was not an offhand statement, but a considered judgment published in a paper presented at the great convocation at the University of Chicago in 1959 celebrating the hundredth anniversary of the publication of *The Origin of Species*. Apparently, none of the distinguished authorities present told Waddington that a tautology does not explain anything. When I want to know how a fish can become a man, I am not enlightened by being told that the organisms that leave the most offspring are the ones that leave the most offspring.

It is not difficult to understand how leading Darwinists were led to formulate natural selection as a tautology. The contemporary neo-Darwinian synthesis grew out of population genetics, a field anchored in mathematics and concerned with demonstrating how rapidly very small mutational advantages could spread in a population. The advantages in question were assumptions in a theorem, not qualities observed in nature, and the mathematicians naturally tended to think of them as "whatever it was that caused the organism and its descendants to produce more offspring than other members of the species." This way of thinking spread to the zoologists and paleontologists, who found it convenient to assume that their guiding theory was simply true by definition. As long as outside critics were not paying attention, the absurdity of the tautology formulation was in no danger of exposure.

What happened to change this situation is that Popper's comment received a great deal of publicity, and creationists and other unfriendly critics began citing it to support their contention that Darwinism is not really a scientific theory. The Darwinists themselves became aware of a dangerous situation, and thereafter critics raising the tautology claim were firmly told that they were simply demonstrating their inability to understand Darwinism. As we shall see in later chapters, however, in practice natural selection continues to be employed in its tautological formulation.

If the concept of natural selection were really only a tautology I could end the chapter at this point, because a piece of empty repetition obviously does not have the power to guide an evolutionary process in its long journey from the first replicating macromolecule to modern human beings. But although natural selection can be formulated as a tautology, and often has been, it can also be formulated in other ways that are not so easily dismissed. We must go on to consider these other possibilities.

Natural Selection as a Deductive Argument

Visitors to the British Natural History Museum will find prominently on sale the museum's handbook on evolution, written by paleontologist Colin Patterson. When he considers the scientific status of Darwinism, Patterson writes that the theory can be presented in the form of a deductive argument, for example:

1. All organisms must reproduce;
2. All organisms exhibit hereditary variations;
3. Hereditary variations differ in their effect on reproduction;
4. Therefore variations with favorable effects on reproduction will succeed, those with unfavorable effects will fail, and organisms will change.

Patterson observes that the theorem establishes only that some natural selection will occur, not that it is a general explanation for evolution. Actually, the theorem does not even establish that organisms will change. The range of hereditary variations may be narrow, and the variations which survive may be just favorable enough to keep the species as it is. Possibly the species would change a great deal more (in the direction of eventual extinction) if the least favored individuals most often succeeded in reproducing their kind. That the effect of natural selection may be to keep a species from changing is not merely a theoretical possibility. As we shall see in Chapter Four, the prevailing characteristic of fossil species is *stasis*—

the absence of change. There are numerous "living fossils" which are much the same today as they were millions of years ago, at least as far as we can determine.

Patterson is not the only evolutionist who thinks of natural selection as a matter of deductive logic, although most who have used this formulation have thought more highly of the theory than he appears to do. For example, origin of life researcher A. G. Cairns-Smith employed the syllogistic formulation (substantially as Darwin himself stated it) to explain how complex organisms can evolve from very simple ones:

> Darwin persuades us that the seemingly purposeful construction of living things can very often, and perhaps always, be attributed to the operation of natural selection. If you have things that are reproducing their kind; *if* there are sometimes random variations, nevertheless, in the offspring; *if* such variations can be inherited; *if* some such variations can sometimes confer an advantage on their owners; *if* there is competition between the reproducing entities;—*if* there is an overproduction so that not all will be able to produce offspring themselves—then these entities will get better at reproducing their kind. Nature acts as a selective breeder in these circumstances: the stock cannot help but improve.

In fact the stock is often highly successful at resisting improvement, often for millions of years, so there must be something wrong with the logic. This time it is the confusion generated by that word "advantage." Advantage in the proper Darwinist sense, as George Gaylord Simpson explained for us, does not mean improvement as humans measure it. Ants and bacteria are just as advantaged as we are, judged by the exclusive criterion of success in reproduction. In any population some individuals will leave more offspring than others, even if the population is not changing or is headed straight for extinction.

Natural Selection as a Scientific Hypothesis

Up to this point we have been disposing of some simple fallacies to clear the field of distractions, but now we get to the important category which deserves our most respectful scrutiny. I am sure that today most evolutionary scientists would insist that Darwinistic natural selection is a scientific hypothesis which has been so thoroughly tested and confirmed by the evidence that it should be accepted by reasonable persons as a presumptively adequate explanation for the evolution of complex life forms. The hypothesis, to be precise, is that natural selection (in combination with mutation) is an innovative evolutionary process capable of producing new kinds of organs and organisms. That brings us to the critical question: what evidence confirms that this hypothesis is true?

Douglas Futuyma has done the best job of marshalling the supporting evidence, and here are the examples he gives of observations that confirm the creative effectiveness of natural selection:

1. Bacteria naturally develop resistance to antibiotics, and insect pests become resistant to insecticides, because of the differential survival of mutant forms possessing the advantage of resistance.

2. In 1898 a severe storm in Massachusetts left hundreds of dead and dying birds in its wake. Someone brought 136 exhausted sparrows to a scientist named Bumpus, I imagine so they could be cared for, but Bumpus was made of sterner stuff and killed the survivors to measure their skeletons. He found that among male sparrows the larger birds had survived more frequently than the smaller ones, even though the size differential was relatively slight.

3. A drought in the Galapagos Islands in 1977 caused a shortage of the small seeds on which finches feed. As a consequence these birds had to eat larger seeds, which they usually ignore. After one generation there had been so much mortality among the smaller finches, who could not easily eat the larger seeds, that the average size of the birds (and especially their beaks) went up appreciably. Futuyma comments: "Very possibly the birds will evolve back to their previous state if the environment goes back to normal,[3] but we can see in this example what would happen if the birds were forced to live in a consistently dry environment: they would evolve a permanent adaptation to whatever kinds of seeds are consistently available. This is natural selection in action, and it is not a matter of chance."

---

[3] In fact this is exactly what happened. The article "Oscillating Selection on Darwin's Finches" by Gibbs and Grant [*Nature*, vol. 327, p. 511, 1987] reports that small adults survived much better than large ones following the wet year 1982-83, completely reversing the trend of 1977-82.

4. The allele (genetic state) responsible for sickle-cell anemia in African populations is also associated with a trait that confers resistance to malaria. Individuals who are totally free of the sickle-cell allele suffer high mortality from malaria, and individuals who inherit the sickle-cell allele from both parents tend to die early from anemia. Chances for survival are greatest when the individual inherits the sickle-cell allele from one parent but not the other, and so the trait is not bred out of the population. Futuyma comments that the example shows not only that natural selection is effective, but also that it is "an uncaring mechanical process."

5. Mice populations have been observed to cease reproducing and become extinct when they are temporarily "flooded" by the spread of a gene which causes sterility in the males.

6. Finally, Futuyma summarizes Kettlewell's famous observations of "industrial melanism" in the peppered moth. When trees were darkened by industrial smoke, dark-colored (melanic) moths became abundant because predators had difficulty seeing them against the trees. When the trees became lighter due to reduced air pollution, the lighter-colored moths had the advantage. Kettlewell's observations showed in detail how the prevailing color of moths changed along with the prevailing color of the trees. Subsequent commentators have observed that the example shows stability as well as cyclical change within a boundary, because the ability of the species to survive in a changing environment is enhanced if it maintains at all times a supply of both light and dark moths. If the light variety had disappeared altogether during the years of dark trees, the species would have been threatened with extinction when the trees lightened.

There are a few other examples in Futuyma's chapter, but I believe they are meant as illustrations to show how Darwinism accounts for certain anomalies like self-sacrificing behavior and the peacock's fan rather than as additional examples of observations confirming the effect of natural selection in producing change. If we take these six examples as the best available observational evidence of natural selection, we can draw two conclusions:

1. There is no reason to doubt that peculiar circumstances can sometimes favor drug-resistant bacteria, or large birds as opposed

to small ones, or dark-colored moths as opposed to light-colored ones. In such circumstances the population of drug-susceptible bacteria, small birds, and light-colored moths may become reduced for some period of time, or as long as the circumstances prevail.

2. None of the "proofs" provides any persuasive reason for believing that natural selection can produce new species, new organs, or other major changes, or even minor changes that are permanent. The sickle-cell anemia case, for example, merely shows that in special circumstances an apparently disadvantageous trait may not be eliminated from the population. That larger birds have an advantage over smaller birds in high winds or droughts has no tendency whatever to prove that similar factors caused birds to come into existence in the first place. Very likely smaller birds have the advantage in other circumstances, which explains why birds are not continually becoming larger.

Pierre Grassé was as unimpressed by this kind of evidence as I am, and he summarized his conclusions at the end of his chapter on evolution and natural selection:

> The "evolution in action" of J. Huxley and other biologists is simply the observation of demographic facts, local fluctuations of genotypes, geographical distributions. Often the species concerned have remained practically unchanged for hundreds of centuries! Fluctuation as a result of circumstances, with prior modification of the genome, does not imply evolution, and we have tangible proof of this in many panchronic species [i.e. living fossils that remain unchanged for millions of years]. . . .

This conclusion seems so obviously correct that it gives rise to another problem. Why do other people, including experts whose intelligence and intellectual integrity I respect, think that evidence of local population fluctuations confirms the hypothesis that natural selection has the capacity to work engineering marvels, to construct wonders like the eye and the wing? Everyone who studies evolution knows that Kettlewell's peppered moth experiment is the classic demonstration of the power of natural selection, and that Darwinists had to wait almost a century to see even this modest confirmation of their central doctrine. Everyone who studies the experiment also knows that it has nothing to do with the origin of any species, or even any variety, because dark and white moths were present throughout the experiment. Only the ratios of one variety to the other changed. How could intelligent people have been so gullible as to imagine that the Kettlewell experiment in any way supported the ambitious claims of Darwinism? To answer that question we need to consider a fourth way in which natural selection can be formulated.

Natural Selection as a Philosophical Necessity

The National Academy of Sciences told the Supreme Court that the most basic characteristic of science is "reliance upon naturalistic explanations," as opposed to "supernatural means inaccessible to human understanding." In the latter, unacceptable category contemporary scientists place not only God, but also any non-material vital force that supposedly drives evolution in the direction of greater complexity, consciousness, or whatever. If science is to have any explanation for biological complexity at all it has to make do with what is left when the unacceptable has been excluded. Natural selection is the best of the remaining alternatives, probably the only alternative.

In this situation some may decide that Darwinism simply *must* be true, and for such persons the purpose of any further investigation will be merely to explain how natural selection works and to solve the mysteries created by apparent anomalies. For them there is no need to test the theory itself, for there is no respectable alternative to test it against. Any persons who say the theory itself is inadequately supported can be vanquished by the question "Darwin's Bulldog" T. H. Huxley used to ask the doubters in Darwin's time: What is your alternative?

I do not think that many scientists would be comfortable accepting Darwinism solely as a philosophical principle, without seeking to find at least some empirical evidence that it is true. But there is an important difference between going to the empirical evidence to test a doubtful theory against some plausible alternative, and going to the evidence to look for confirmation of the only theory that one is willing to tolerate. We have already seen that distinguished scientists have accepted uncritically the questionable analogy between natural and artificial selection, and that they have often been undisturbed by the fallacies of the "tautology" and "deductive logic" formulations. Such illogic survived and reproduced itself for the same reason that an apparently incompetent species sometimes avoids extinction; there was no effective competition in its ecological niche.

If positive confirmation of the creative potency of natural selection is not required, there is little danger that the theory will be disproved by negative evidence. Darwinists have evolved an array of subsidiary concepts capable of furnishing a plausible explanation for just about any conceivable eventuality. For example, the living fossils, which have remained basically unchanged for millions of years while their cousins were supposedly evolving into more advanced creatures like human beings, are no embarrassment to Darwinists. They failed to evolve because the necessary mutations didn't arrive, or because of "developmental constraints," or because they were already adequately adapted to their environment. In short, they didn't evolve because they didn't evolve.

Some animals give warning signals at the approach of predators, apparently reducing their own safety for the benefit of others in the herd. How does natural selection encourage the evolution of a trait for self-sacrifice? Some Darwinists attribute the apparent anomaly to "group selection." Human nations benefit if they contain individuals willing to die in battle for their country, and likewise animal groups containing self-sacrificing individuals may have an advantage over groups composed exclusively of selfish individuals.

Other Darwinists are scornful of group selection and prefer to explain altruism on the basis of "kinship selection." By sacrificing itself to preserve its offspring or near relations an individual promotes the survival of its genes. Selection may thus operate at the genetic level to encourage the perpetuation of genetic combinations that produce individuals capable of altruistic behavior. By moving the focus of selection either up (to the group level) or down (to the genetic level), Darwinists can easily account for traits that seem to contradict the selection hypothesis at the level of individual organisms.

Potentially the most powerful explanatory tool in the entire Darwinist armory is *pleiotropy*, the fact that a single gene has multiple effects. This means that any mutation which affects one functional characteristic is likely to change other features as well, and whether or not it is advantageous depends upon the net effect. Characteristics which on their face appear to be maladaptive may therefore be presumed to be linked genetically to more favorable characteristics, and natural selection can be credited with preserving the package.

I am not implying that there is anything inherently unreasonable in invoking pleiotropy, or kinship selection, or developmental constraints to explain why apparent anomalies are not necessarily inconsistent with Darwinism. If we assume that Darwinism is basically true then it is perfectly reasonable to adjust the theory as necessary to make it conform to the observed facts. The problem is that the adjusting devices are so flexible that in combination they make it difficult to conceive of a way to test the claims of Darwinism empirically. Apparently maladaptive features can be attributed to pleiotropy, or to our inability to perceive the advantage that may be there, or when all else fails simply to "chance." Darwin wrote that "If it could be proved that any part of the structure of any one species had been formed for the exclusive good of another species, it would annihilate my theory, for such could not have been produced through natural selection." But this was the same Darwin who insisted that he had never claimed that natural selection was the exclusive mechanism of evolution.

One important subsidiary concept—sexual selection—illustrates the skill of Darwinists at incorporating recalcitrant examples into their theory. Sexual selection is a relatively minor component in Darwinist theory today, but to Darwin it was almost as important as natural selection itself. (Darwin's second classic, *The Descent of Man*, is mainly a treatise on sexual selection.) The most famous example of sexual selection is the peacock's gaudy fan, which is obviously an encumbrance when a peacock wants to escape a predator. The fan is stimulating to peahens, however, and so its possession increases the peacock's prospects for producing progeny even though it decreases his life expectancy.

The explanation so far is reasonable, even delightful, but what I find intriguing is that Darwinists are not troubled by the unfitness of the peahen's sexual taste. Why would natural selection, which supposedly formed all birds from lowly predecessors, produce a species whose females lust for males with life-threatening decorations? The peahen ought to have developed a preference for males with sharp talons and mighty wings. Perhaps the taste for fans is associated genetically with some absolutely vital trait like strong egg shells, but then why and how did natural selection encourage such an absurd genetic linkage? Nevertheless, Douglas Futuyma boldly proclaims the peacock as a problem not for Darwinists but for creationists:

> Do the creation scientists really suppose their Creator saw fit to create a bird that couldn't reproduce without six feet of bulky feathers that make it easy prey for leopards?

I don't know what creation-scientists may suppose, but it seems to me that the peacock and peahen are just the kind of creatures a whimsical Creator might favor, but that an "uncaring mechanical process" like natural selection would never permit to develop.

What we are seeing in Futuyma's comment about the peacock is the debating principle that the best defense is a good offense, but we are also seeing the influence of philosophical preconception in blinding an intelligent Darwinist to the existence of a counterexample. Julian Huxley once wrote that "Improbability is to be *expected* as a result of natural selection; and we have the paradox that an exceedingly high apparent improbability in its products can be taken as evidence for the high degree of its efficacy." On that basis the theory has nothing to fear from the evidence.

Natural selection is the most famous element in Darwinism, but it is not necessarily the most important element. Selection merely preserves or destroys something that already exists. Mutation has to provide the favorable innovations before natural selection can retain and encourage them. That brings us to our next subject, which requires a separate chapter.

Chapter Three

Mutations Great and Small

"EVOLUTION" IS A concept broad enough to encompass just about any alternative to instantaneous creation, and so it is not surprising that thinkers have speculated about evolution ever since ancient times. Charles Darwin's unique contribution was to describe a plausible mechanism by which the necessary transformations could occur, a mechanism that did not require divine guidance, mysterious vital forces, or any other causes not presently operating in the world. Darwin was particularly anxious to avoid the need for any "saltations"—sudden leaps by which a new type of organism appears in a single generation. Saltations (or systemic macromutations, as they are often called today) are believed to be theoretically impossible by most scientists, and for good reason. Living creatures are extremely intricate assemblies of interrelated parts, and the parts themselves are also complex. It is impossible to imagine how the parts could change in unison as a result of chance mutation.

In a word (Darwin's word), a saltation is equivalent to a miracle. At the extreme, saltationism is virtually indistinguishable from special creation. If a snake's egg were to hatch and a mouse emerge, we could with equal justice classify the event as an instance of evolution or creation. Even the sudden appearance of a single complex organ, like an eye or wing, would imply supernatural intervention. Darwin emphatically rejected any evolutionary theory of this sort, writing to Charles Lyell that

> If I were convinced that I required such additions to the theory of natural selection, I would reject it as rubbish. . . . I would give nothing for the theory of natural selection, if it requires miraculous additions at any one stage of descent.

Darwin aimed to do for biology what Lyell had done for geology: explain great changes on uniformitarian and naturalistic principles, meaning the gradual operation over long periods of time of familiar natural forces that we can still see operating in the present. He understood that the distinctive feature of his theory was its uncompromising philosophical materialism, which made it truly scientific in the sense that it did not invoke any mystical or supernatural forces that are inaccessible to scientific investigation. To achieve a fully materialistic theory Darwin had to explain every complex characteristic or major transformation as the cumulative product of a great many tiny steps. In his own eloquent words:

> Natural selection can act only by the preservation and accumulation of infinitesimally small inherited modifications, each profitable to the preserved being; and as modern geology has almost banished such views as the excavation of a great valley by a single diluvial wave, so will natural selection, if it be a true principle, banish the belief of the continued creation of new organic beings, or of any great and sudden modification in their structure.

T. H. Huxley protested against this dogmatic gradualism from the start, warning Darwin in a famous letter that "You have loaded yourself with an unnecessary difficulty in adopting *natura non facit saltum* so unreservedly." The difficulty was hardly unnecessary, given Darwin's purpose, but it was real enough. In the long term the biggest problem was the fossil record, which did not provide evidence of the many transitional forms that Darwin's theory required to have existed. Darwin made the obvious response, arguing that the evidence was lacking because the fossil record was incomplete. This was a reasonable possibility at the time, and conveniently safe from disproof; we shall return to it in the next chapter.

The more pressing difficulty was theoretical. Many organs require an intricate combination of complex parts to perform their functions. The eye and the wing are the most common illustrations, but it would be misleading to give the impression that either is a special case; human and animal bodies are literally packed with similar marvels. How can such things be built up by "infinitesimally small inherited variations, *each* profitable to the preserved being?" The first step towards a new function—such as vision or ability to fly—would not necessarily provide any advantage unless the other parts required for the function appeared at the same time. As an analogy, imagine a medieval alchemist producing by chance a silicon microchip; in the absence of a supporting computer technology the prodigious invention would be useless and he would throw it away.

Stephen Jay Gould asked himself "the excellent question, What good is 5 per cent of an eye?," and speculated that the first eye parts might have been useful for something other than sight. Richard Dawkins responded that

> An ancient animal with 5 per cent of an eye might indeed have used it for something other than sight, but it seems to me as likely that it used it for 5 per cent vision. And actually I don't think it is an excellent question. Vision that is 5 per cent as good as yours or mine is very much worth having in comparison with no vision at all. So is 1 per cent vision better than total blindness. And 6 per cent is better than 5, 7 per cent better than 6, and so on up the gradual, continuous series.

The fallacy in that argument is that "5 per cent of an eye" is not the same thing as "5 per cent of normal vision." For an animal to have any useful vision at all, many complex parts must be working together. Even a complete eye is useless unless it belongs to a creature with the mental and neural capacity to make use of the information by doing something that furthers survival or reproduction.

What we have to imagine is a chance mutation that provides this complex capacity all at once, at a level of utility sufficient to give the creature an advantage in producing offspring.

Dawkins went on to restate Darwin's answer to the eye conundrum, pointing out that there is a plausible series of intermediate eye-designs among living animals. Some single-celled animals have a light-sensitive spot with a little pigment screen behind it, and in some many-celled animals a similar arrangement is set in a cup, which gives improved direction-finding capability. The ancient nautilus has a pinhole eye with no lens, the squid's eye adds the lens, and so on. None of these different types of eyes are thought to have evolved from any of the others, however, because they involve different types of structures rather than a series of similar structures growing in complexity.

If the eye evolved at all, it evolved many times. Ernst Mayr writes that the eye must have evolved independently at least 40 times, a circumstance which suggests to him that "a highly complicated organ can evolve repeatedly and convergently when advantageous, provided such evolution is at all probable." But then why did the many primitive eye forms that are still with us never evolve into more advanced forms? Dawkins admits to being baffled by the nautilus, which in its hundreds of millions of years of existence has never evolved a lens for its eye despite having a retina that is "practically crying out for (this) particular simple change."[1]

The wing, which exists in quite distinct forms in insects, birds, and bats, is the other most frequently cited puzzle. Would the first "infinitesimally small inherited modification" in the direction of wing construction confer a selective advantage? Dawkins thinks that it would, because even a small flap or web might help a small creature to jump farther, or save it from breaking its neck in a fall. Eventually such a proto-wing might develop to a point where the creature would begin gliding, and by further gradual improvements it would become capable of genuine flight. What this imaginative scenario neglects is that forelimbs evolving into wings would

---

[1] Before leaving the subject of the eye, I should add that Darwinists cite *imperfections* in the eye as evidence that it was not designed by an omniscient creator. According to Dawkins, the photocells are "wired backwards," and "any tidy-minded engineer" would not have been so sloppy.

probably become awkward for climbing or grasping long before they became very useful for gliding, thus placing the hypothetical intermediate creature at a serious disadvantage.

There is a good skeptical discussion of the bird wing problem in chapter 9 of Denton's *Evolution: A Theory in Crisis*. Denton describes the exquisitely functional avian feather, with its interlocking hooks and other intricate features that make it suitable for flight and quite distinct from any form of feather used only for warmth. Bird feathers must have evolved from reptilian scales if Darwinism is true, but once again the intermediates are hard to imagine. Still more difficult a problem is presented by the distinctive avian lung, which is quite different in structure than that of any conceivable evolutionary ancestor. According to Denton,

> Just how such a different respiratory system could have evolved gradually from the standard vertebrate design is fantastically difficult to envisage, especially bearing in mind that the maintenance of respiratory function is absolutely vital to the life of an organism to the extent that the slightest malfunction leads to death within minutes. Just as the feather cannot function as an organ of flight until the hooks and barbules are coadapted to fit together perfectly, so the avian lung cannot function as an organ of respiration until the parabronchi system which permeates it and the air sac system which guarantees the parabronchi their air supply are both highly developed and able to function together in a perfectly integrated manner.

Whether one finds the gradualist scenarios for the development of complex systems plausible involves an element of subjective judgment. It is a matter of objective fact, however, that these scenarios are speculation. Bird and bat wings appear in the fossil record already developed, and no one has ever confirmed by experiment that the gradual evolution of wings and eyes is possible. This absence of historical or experimental confirmation is presumably what Gould had in mind when he wrote that "These tales, in the 'just-so' tradition of evolutionary natural history, do not prove anything." Are we dealing here with science or with rationalist versions of Kipling's fables?

Darwin wrote that "If it could be demonstrated that any complex organ existed which could not possibly have been formed by numerous, successive, slight modifications, my theory would absolutely break down." One particularly eminent scientist of the mid-twentieth century who concluded that it had absolutely broken down was the German-American geneticist, Professor Richard Goldschmidt of the University of California at Berkeley. Goldschmidt issued a famous challenge to the neo-Darwinists, listing a series of complex structures from mammalian hair to hemoglobin that he thought could not have been produced by the accumulation and selection of small mutations. Like Pierre Grassé, Goldschmidt concluded that Darwinian evolution could account for no more than variations within the species boundary; unlike Grassé, he thought that evolution beyond that point must have occurred in single jumps through macromutations. He conceded that large-scale mutations would in almost all cases produce hopelessly maladapted monsters, but he thought that on rare occasions a lucky accident might produce a "hopeful monster," a member of a new species with the capacity to survive and propagate (but with what mate?).

The Darwinists met this fantastic suggestion with savage ridicule. As Goldschmidt put it, "This time I was not only crazy but almost a criminal." Gould has even compared the treatment accorded Goldschmidt in Darwinist circles with the daily "Two Minute Hate" directed at "Emmanuel Goldstein, enemy of the people" in George Orwell's novel *1984*. The venom is explained by the emotional attachment Darwinists have to their theory, but the ridicule had a sound scientific basis. If Goldschmidt really meant that all the complex interrelated parts of an animal could be reformed together in a single generation by a systemic macromutation, he was postulating a virtual miracle that had no basis either in genetic theory or in experimental evidence. Mutations are thought to stem from random errors in copying the commands of the DNA's genetic code. To suppose that such a random event could reconstruct even a single complex organ like a liver or kidney is about as reasonable as to suppose that an improved watch can be designed by throwing an old one against a wall. Adaptive macromutations are impossible, say the Darwinists, especially if required in any quantity, and so all those complex organs must have evolved—many times independently—by the selective accumulation of micromutations over a long period of time.

But now we must deal with another fallacy, and a supremely important one. That evolution by macromutation is impossible does not prove that evolution by micromutation is probable, or even possible. It is likely that Darwinist gradualism is statistically just as unlikely as Goldschmidt's saltationism, once we give adequate attention to all the necessary elements. The advantageous micromutations postulated by neo-Darwinist genetics are tiny, usually too small to be noticed. This premise is important because, in the words of Richard Dawkins, "virtually all the mutations studied in genetics laboratories—which are pretty macro because otherwise geneticists wouldn't notice them—are deleterious to the animals possessing them." But if the necessary mutations are too small to be seen, there will have to be a great many of them (millions?) of the right type coming along when they are needed to carry on the long-term project of producing a complex organ.

The probability of Darwinist evolution depends upon the quantity of favorable micromutations required to create complex organs and organisms, the frequency with which such favorable micromutations occur just where and when they are needed, the efficacy of natural selection in preserving the slight improvements with sufficient consistency to permit the benefits to accumulate, and the time allowed by the fossil record for all this to have happened. Unless we can make calculations taking all these factors into account, we have no way of knowing whether evolution by micromutation is more or less improbable than evolution by macromutation.

Some mathematicians did try to make the calculations, and the result was a rather acrimonious confrontation between themselves and some of the leading Darwinists at the Wistar Institute in Philadelphia in 1967. The report of the exchange is fascinating, not just because of the substance of the mathematical challenge, but even more because of the logic of the Darwinist response. For example, the mathematician D. S. Ulam argued that it was highly improbable that the eye could have evolved by the accumulation of small mutations, because the number of mutations would have to be so large and the time available was not nearly long enough for them to appear. Sir Peter Medawar and C. H. Waddington responded that Ulam was doing his science backwards; the fact was that the eye *had* evolved and therefore the mathematical difficulties must be only apparent. Ernst Mayr observed that Ulam's calculations were based on assumptions that might be unfounded, and concluded that

"Somehow or other by adjusting these figures we will come out all right. We are comforted by the fact that evolution has occurred."

The Darwinists were trying to be reasonable, but it was as if Ulam had presented equations proving that gravity is too weak a force to prevent us all from floating off into space. Darwinism to them was not a theory open to refutation but a fact to be accounted for, at least until the mathematicians could produce an acceptable alternative. The discussion became particularly heated after a French mathematician named Schützenberger concluded that "there is a considerable gap in the neo-Darwinian theory of evolution, and we believe this gap to be of such a nature that it cannot be bridged within the current conception of biology." C. H. Waddington thought he saw where this reasoning was headed, and retorted that "Your argument is simply that life must have come about by special creation." Schützenberger (and anonymous voices from the audience) shouted "No!," but in fact the mathematicians did not present an alternative.

The difficulties with both the micromutational and macromutational theories are so great that we might expect to see some effort being made to come up with a middle ground that minimizes the disadvantages of both extremes. Stephen Jay Gould attempted something of the sort, both in his 1980 scientific paper proposing a "new and general theory," and in his popular article "The Return of the Hopeful Monster." Gould tried to rehabilitate Goldschmidt while domesticating his monster. Goldschmidt did not really mean that "new species arise all at once, fully formed, by a fortunate macromutation," Gould explained, and what he did mean can be reconciled with "the essence of Darwinism."

> Suppose that a discontinuous change in adult form arises from a small genetic alteration. Problems of discordance with other members of the species do not arise, and the large, favorable variant can spread through a population in Darwinian fashion. Suppose also that this large change does not produce a perfected form all at once, but rather serves as a "key" adaptation to shift its possessor toward a new mode of life. Continued success in this new mode may require a large set of collateral alterations, morphological and behavioral; these may arise by a more traditional, gradual route once the key adaptation forces a profound shift in selective pressures.

We have to do all this supposing, according to Gould, because it is just too hard to "invent a reasonable sequence of intermediate forms—that is, viable, functioning organisms -between ancestors and descendants in major structural transitions." In the end we will have to accept "many cases of discontinuous transition in macro-evolution." The kind of small genetic alteration which Gould had in mind (and said Goldschmidt had in mind) was a mutation in the genes regulating embryonic development, on the theory that "small changes early in embryology accumulate through growth to yield profound differences among adults." Indeed they must do so, because otherwise Gould could not see any way that major evolutionary transitions could have been accomplished.

Gould published a major article in the scientific journal *Paleobiology* which expressed his endorsement of Goldschmidt even more explicitly, and in which he pronounced the effective death of the neo-Darwinian synthesis. In place of the dead orthodoxy he hailed as "the epitome and foundation of emerging views on speciation" a passage by Goldschmidt which insisted that "neo-Darwinian evolution . . . is a process which leads to diversification strictly within the species. . . . The decisive step in evolution, the first step towards macroevolution, the step from one species to another, requires another evolutionary method than the sheer accumulation of micro-mutations." With respect to the evolution of complex organs, Gould disavowed reliance on "saltational origin of entire new designs," but proposed instead "a potential saltational origin for the essential features of key adaptations." In short, he tried to split the difference between Darwinism and Goldschmidtism.

And so the hopeful monster returned, but its hopes were soon disappointed once again. Ernst Mayr, the most prestigious of living neo-Darwinists, wrote that Gould had entirely misrepresented Goldschmidt's theory in denying that Goldschmidt advocated impossible, single-generation systemic macromutations. "Actually, this is what Goldschmidt repeatedly claimed. For instance, he cited with approval Schindewolf's[2] suggestion that the first bird hatched out of a reptilian egg. . . ." Mayr thought that some mutations with large

---

[2] Otto Schindewolf was a prominent paleontologist whom we will encounter again in the next chapter.

scale effects might be possible,[3] but he could find no evidence that any great number of them had occurred and he saw no need to invoke them because he considered the mechanisms of neo-Darwinism capable of explaining the emergence of evolutionary novelties.

Richard Dawkins wrote scornfully of Goldschmidt in *The Blind Watchmaker*, and criticized Gould for trying to rehabilitate him. For Dawkins, "Goldschmidt's problem . . . turns out to be no problem at all," because there is no real difficulty in accounting for the development of complex structures by gradualistic evolution. What Dawkins seems to mean by this assertion is that the step-by-step evolution of complex adaptive systems is a conceptual possibility, not that there is some way to prove that it actually happens. He uses the bat, with its marvelous sonar-like echolocation system that so resembles the product of an advanced technological society, as the paradigm example of how natural selection can explain the development of a complex system that would otherwise be taken as evidence for the existence of a "watchmaker" creator. Dawkins is right to argue that if Darwinist evolution can craft a bat it can make just about anything, but what he neglects to do is to prove that Darwinist evolution can do anything of the kind. It is *conceivable* that bat sonar evolved by some step-by-step process, in which the first hint of an ability to locate by echo was of such value to its possessor that everything else had to follow, but how do we know that such a thing ever happened, or could have happened?

Despite his generally rigid adherence to Darwinist gradualism, even Dawkins finds it impossible to get along without what might be called modest macromutations, meaning mutations that "although they may be large in the magnitude of their effects, turn out not to be large in terms of their complexity." He uses as an example snakes, some contemporary examples of which have more vertebrae than their presumed ancestors. The number of vertebrae has to be

---

[3] The debate over macromutations has mainly concerned the animal kingdom, but it is well known that a special kind of macromutation, known as polyploidy, can produce new plant species. This phenomenon, which involves the doubling of chromosome numbers in cell division applies only to hermaphrodite species capable of self-fertilization. As a result it is important only for plants, although not entirely absent from the animal kingdom. In any case, polyploidy would not explain the creation of complex adaptive structures like wings and eyes.

changed in whole units, and to accomplish this "you need to do more than just shove in an extra bone," because each vertebra has associated with it a set of nerves, blood vessels, muscles, and so on. These complicated parts would all have to appear together for the extra vertebrae to make any biological sense, but "it is easy to believe that individual snakes with half a dozen more vertebrae than their parents could have arisen in a single mutational step." This is easy to believe, according to Dawkins, because the mutation only adds more of what was already there, and because the change only appears to be macromutational when we look at the adult. At the embryonic level, such changes "turn out to be micromutations, in the sense that only a small change in the embryonic instructions had a large apparent effect in the adult."

Gould supposes what he has to suppose, and Dawkins finds it easy to believe what he wants to believe, but supposing and believing are not enough to make a scientific explanation. Is there any way to confirm the hypothesis that mutations in the genes which regulate embryonic development might provide whatever is needed to get evolution over the unbridgeable gaps? Creatures that look very different as adults are sometimes much more alike at the early embryonic stages, and so there is a certain plausibility to the notion that a simple but basic change in the genetic program regulating development could induce an embryo to develop in an unusual direction. In principle, this is the kind of change we might imagine human genetic engineers to be capable of directing one day, if this branch of science continues to advance in the future as it has in the recent past.

Suppose that, following a massive research program, scientists succeed in altering the genetic program of a fish embryo so that it develops as an amphibian. Would this hypothetical triumph of genetic engineering confirm that amphibians actually evolved, or at least could have evolved, in similar fashion?

No it wouldn't, because Gould and the others who postulate developmental macromutations are talking about *random* changes, not changes elaborately planned by human (or divine) intelligence. A random change in the program governing my word processor could easily transform this chapter into unintelligible gibberish, but it would not translate the chapter into a foreign language, or produce a coherent chapter about something else. What the proponents of developmental macromutations need to establish is not merely that there is an alterable genetic program governing development, but that important evolutionary innovations can be produced by random changes in the genetic instructions.

The prevailing assumption in evolutionary science seems to be that speculative possibilities, without experimental confirmation, are all that is really necessary. The principle at work is the same one that Waddington, Medawar, and Mayr invoked when challenged by the mathematicians. Nature must have provided whatever evolution had to have, because otherwise evolution wouldn't have happened. It follows that if evolution required macromutations then macromutations must be possible, or if macromutations are impossible then evolution must not have required them. The theory itself provides whatever supporting evidence is essential.

If the Darwinists are at all uncomfortable with this situation (actually, most of them don't seem to be), the anti-Darwinists are in no better shape. The great geneticist Goldschmidt was reduced to endorsing a genetic impossibility, and the great zoologist Grassé could do no better than to suggest that evolving species somehow acquire a new store of genetic information due to obscure "internal factors" involving "a phenomenon whose equivalent cannot be seen in the creatures living at the present time (either because it is not there or because we are unable to see it)." Grassé was all too aware that such talk "arouses the suspicions of many biologists . . . [because] it conjures up visions of the ghost of vitalism or of some mystical power which guides the destiny of living things. . . ." He repeatedly denied that he had anything of the sort in mind, but suspicions of vitalism once aroused are not conjured away by bare denials.

We can see from these examples why neo-Darwinism retains its status as textbook orthodoxy despite all the difficulties and even the imputations of moribundity. If neo-Darwinist gradualism were abandoned as incapable of explaining macroevolutionary leaps and the origin of complex organs, most biologists would still believe in evolution (Goldschmidt and Grassé never doubted that evolution had occurred), but they would have no *theory* of evolution. Materialist scientists are full of scorn for creationists who invoke an invisible creator who employed supernatural powers that cannot be observed operating in our own times. If evolutionary science must also rely upon mystical guiding forces or upon genetically impossible transformations, a philosophical materialist like Charles Darwin would call it rubbish.

Until now I have avoided discussing the fossil evidence in order to concentrate on the theoretical and experimental difficulties that surround the reigning neo-Darwinist synthesis. But evolution is at bottom about history; it aims to tell us what happened in the past. On that subject the fossils are our most direct evidence, and it is to them that we turn next.

Chapter Four

The Fossil Problem

TODAY IT IS widely assumed that the existence of fossil remains of numerous extinct species necessarily implies evolution, and most people are unaware that Darwin's most formidable opponents were not clergymen, but fossil experts. In the early nineteenth century the prevailing geological theory was the "catastrophism" advocated by the great French scientist Cuvier, the father of paleontology. Cuvier believed that the geological record showed a pattern of catastrophic events involving mass extinctions, which were followed by periods of creation in which new forms of life appeared without any trace of evolutionary development.

In Darwin's time, Cuvier's catastrophism was being supplanted by the uniformitarian geology advocated by Darwin's older friend Charles Lyell, who explained spectacular natural features as the result not of sudden cataclysms, but rather the slow working over immense time of everyday forces. In retrospect, an evolutionary theory of the Darwinian kind seems almost an inevitable extension

45

of Lyell's logic, but Lyell himself had great difficulty accepting biological evolution, as did many other persons who were familiar with the evidence.

Each of the divisions of the biological world (kingdoms, phyla, classes, orders), it was noted, conformed to a basic structural plan, with very few intermediate types. Where were the links between these discontinuous groups? The absence of transitional intermediates was troubling even to Darwin's loyal supporter T. H. Huxley, who warned Darwin repeatedly in private that a theory consistent with the evidence would have to allow for some big jumps.

Darwin posed the question himself, asking

why, if species have descended from other species by insensibly fine gradations, do we not everywhere see innumerable transitional forms? Why is not all nature in confusion instead of the species being, as we see them, well defined?

He answered with a theory of extinction which was the logical counterpart of "the survival of the fittest." The appearance of an improved form implies a disadvantage for its parent form. Thus, "if we look at each species as descended from some other unknown form, both the parent and all the transitional varieties will generally have been exterminated by the very process of formation and perfection of the new form." This extermination-by-obsolescence implies that *appearances* will be against a theory of evolution in our living world, because we see distinct, stable species (and larger groupings), with only rare intermediate forms. The links between the discontinuous groups that once existed have vanished due to maladaptation.

But what if the necessary links are missing not only from the world of the present, but from the fossil record of the past as well? Darwin acknowledged that his theory implied that "the number of intermediate and transitional links, between all living and extinct species, must have been inconceivably great." One might therefore suppose that geologists would be continually uncovering fossil evidence of transitional forms. This, however, was clearly not the case. What geologists did discover was species, and groups of species, which appeared suddenly rather than at the end of a chain of evolutionary links. Darwin conceded that the state of the fossil evidence was "the most obvious and gravest objection which can be urged against my theory," and that it accounted for the fact that "all the most eminent paleontologists . . . and all our greatest geologists . . . have unanimously, often vehemently, maintained the immutability of species."

Darwin argued eloquently that the fossil problem, although concededly serious, was not fatal to his theory. His main point was that the fossil record is extremely imperfect. Fossils are preserved only in special circumstances, and thus the various fossil beds of the world probably reflect not a continuous record but rather pictures of relatively brief periods separated from each other by wide intervals of time. Additionally, we might fail to recognize ancestor-descendant relationships in the fossils even if they were present. Unless we had all the intervening links to show the connections between them, the two forms might appear entirely distinct to our eyes. At times Darwin even seemed to be implying that the absence of transitionals was itself a proof of the inadequacy of the record, as it would be if one had *a priori* knowledge that his theory was true:

> I do not pretend that I should ever have suspected how poor a record of the mutations of life, the best preserved geological section presented, had not the difficulty of our not discovering innumerable transitional links between the species which appeared at the commencement and close of each formation, pressed so hardly on my theory.

Darwin did as well with the fossil problem as the discouraging facts allowed, but to some questions he had to respond frankly that "I can give no satisfactory answer," and there is a hint of desperation in his writing at times, as in the following sentence: "Nature may almost be said to have guarded against the frequent discovery of her transitional or linking forms." But Darwin never lost faith in his theory; the only puzzle was how to account for the plainly misleading aspects of the fossil record.

At this point I ask the reader to stop with me for a moment and consider what an unbiased person ought to have thought about the controversy over evolution in the period immediately following the publication of *The Origin of Species*. Opposition to Darwin's theory could hardly be attributed to religious prejudice when the skeptics included the leading paleontologists and geologists of the day. Darwin's defense of the theory against the fossil evidence was not unreasonable, but the point is, it was a *defense*. Very possibly the fossil beds are mere snapshots of moments in geological time, with sufficient time and space between them for a lot of evolution to be going on in the gaps. Still, it is one thing to say that there are gaps, and quite another thing to claim the right to fill the gaps with the evidence required to support one's theory. Darwin's arguments could establish at most that the fossil problem was not fatal; they could not turn the absence of confirming evidence into an asset.

There was a way to test the theory by fossil evidence, however, if Darwin and his followers had wanted a test. Darwin was emphatic that the number of transitional intermediates must have been immense, even "inconceivable." Perhaps evidence of their existence was missing because in 1859 only a small part of the world's fossil beds had been searched, and because the explorers had not known what to look for. Once paleontologists accepted Darwinism as a working hypothesis, however, and explored many new fossil beds in an effort to confirm the theory, this situation ought to change. In time the fossil record could be expected to look very different, and very much more Darwinian.

The test would not be fair to the skeptics, however, unless it was also possible for the theory to fail. Imagine, for example, that belief in Darwin's theory were to sweep through the scientific world with such irresistible power that it very quickly became an orthodoxy. Suppose that the tide was so irresistible that even the most prestigious of scientists—Harvard's Louis Agassiz, for example—became an instant has-been for failing to join the movement. Suppose that paleontologists became so committed to the new way of thinking that fossil studies were published only if they supported the theory, and were discarded as failures if they showed an absence of evolutionary change. As we shall see, that is what happened. Darwinism apparently passed the fossil test, but only because it was not allowed to fail.

Darwin's theory predicted not merely that fossil transitionals would be found; it implied that a truly complete fossil record would be mostly transitionals, and that what we think of as fixed species would be revealed as mere arbitrary viewpoints in a process of continual change. Darwinism also implied an important prediction about extinction, that necessary corollary of the struggle for existence. Darwin recognized that his theory required a pattern of extinction even more gradual than the pattern of evolutionary emergence:

> The old notion of all the inhabitants of the earth having been swept away at successive periods by catastrophes, is very generally given up, even by those geologists . . . whose general views would naturally lead them to this conclusion. . . . There is reason to believe that the complete extinction of the species of a group is generally a slower process than their production: if the appearance and disappearance of a group of species be represented, as before, by a vertical line of varying thickness, the line is found to taper more gradually at its upper end, which marks the progress of extermination, than in its lower end, which marks the first appearance and increase in numbers of the species. In some cases, however, the extermination of whole groups of beings, as of ammonites towards the close of the secondary period, has been wonderfully sudden.

Continual, gradual extinctions are a necessary consequence of the assumption that ancestor species are constantly being supplanted by better adapted descendants. Suppose, however, that it were shown that a substantial proportion of extinctions have occurred in the course of a few global catastrophes, such as might be caused by a comet hitting the earth or some sudden change in temperature. In such catastrophes survival would not necessarily have been related to fitness in more normal circumstances, and might have been entirely at random. Darwinism could therefore be tested not only by searching for transitional species in newly discovered fossil beds, but also by studying the pattern of extinctions to measure the importance of catastrophes.

Evolution triumphed during Darwin's lifetime, although his opposition to saltations remained controversial in scientific circles for a long time to come. The discovery of *Archaeopteryx*—an ancient bird with some strikingly reptilian features—was enough fossil confirmation in itself to satisfy many. Thereafter it was one apparent fossil success after another, with reports of human ancestors, ancient mammal-like reptiles, a good sequence in the horse line, and so on. Paleontology joined the neo-Darwinian synthesis in the work of George Gaylord Simpson, who declared that Darwin had been confirmed by the fossils (a declaration that was communicated to generations of biology students as fact). What Stephen Jay Gould described in 1980 as "the most sophisticated of modern American textbooks for introductory biology" endorsed the synthetic theory on the basis of fossil evidence:

[Can] more extensive evolutionary change, macroevolution, be explained as an outcome of these microevolutionary shifts? Did birds really arise from reptiles by an accumulation of gene substitutions of the kind illustrated by the raspberry eye-color gene?

The answer is that it is entirely plausible, and no one has come up with a better explanation. . . .The fossil record suggests that macroevolution is indeed gradual, paced at a rate that leads to the conclusion that it is based on hundreds or thousands of gene substitutions no different in kind from the ones examined in our case histories.

But that last sentence is false, and has long been known to paleontologists to be false.

The fossil record was revisited in the 1970s in works by Stephen Jay Gould, Niles Eldredge, and Steven Stanley. Gould and Eldredge proposed a new theory they called "punctuated equilibrium" ("punk eek" to the irreverent), to deal with an embarrassing fact: the fossil record today on the whole looks very much as it did in 1859, despite the fact that an enormous amount of fossil hunting has gone on in the intervening years. In the words of Gould:

The history of most fossil species includes two features particularly inconsistent with gradualism:

1. Stasis. Most species exhibit no directional change during their tenure on earth. They appear in the fossil record looking pretty much the same as when they disappear; morphological change is usually limited and directionless.
2. Sudden appearance. In any local area, a species does not arise gradually by the steady transformation of its ancestors; it appears all at once and "fully formed."

In short, if evolution means the gradual change of one kind of organism into another kind, the outstanding characteristic of the fossil record is the absence of evidence for evolution. Darwinists can always explain away the sudden appearance of new species by saying that the transitional intermediates were for some reason not fossilized. But stasis—the consistent absence of fundamental directional change—is positively documented. It is also the norm and not the exception.

According to Steven Stanley, the Bighorn Basin in Wyoming contains a continuous local record of fossil deposits for about five million years, during an early period in the age of mammals. Because this record is so complete, paleontologists assumed that certain populations of the basin could be linked together to illustrate continuous evolution. On the contrary, species that were once thought to have turned into others turn out to overlap in time with their alleged descendants, and "the fossil record does not convincingly document a single transition from one species to another." In addition, species remain fundamentally unchanged for an average of more than one million years before disappearing from the record. Stanley uses the example of the bat and the whale, which are supposed to have evolved from a common mammalian ancestor in little more than ten million years, to illustrate the insuperable problem that fossil stasis poses for Darwinian gradualism:

> Let us suppose that we wish, hypothetically, to form a bat or a whale . . . [by a] process of gradual transformation of established species. If an average chronospecies lasts nearly a million years, or even longer, and we have at our disposal only ten million years, then we have only ten or fifteen chronospecies[1] to align, end-to-end, to form a continuous lineage connecting our primitive little mammal with a bat or a whale. This is clearly preposterous. Chronospecies, by definition, grade into each other, and each one encompasses very little change. A chain of ten or fifteen of these might move us from one small rodentlike form to a slightly different one, perhaps representing a new genus, but not to a bat or a whale!

To provide more rapid change Stanley relies partly upon the so far untestable theory that random mutations in the "regulatory genes" might alter the program for embryonic development suffi-

---

[1] In the living world, species are separate reproductive communities, which do not interbreed. Because we cannot determine the breeding capabilities of creatures known only by fossils, these have to be assigned to species by their visible characteristics. A "chronospecies" is a segment of a fossil lineage judged to have evolved so little in observable characteristics that it remained a single species.

ciently to produce a new form in a single generation. Whether or not macromutations are involved, the most important concept of evolution by punctuated equilibrium, as developed by Gould and Eldredge, is that speciation (the formation of new species) occurs rapidly,[2] and in small groups which are isolated on the periphery of the geographical area occupied by the ancestral species. Selective pressures might be particularly intense in an area where members of the species are just barely able to survive, and favorable variations could spread relatively quickly through a small, isolated population. By this means a new species might arise in the peripheral area without leaving fossil evidence. Because fossils are mostly derived from large, central populations, a new species would appear suddenly in the fossil record following its migration into the central area of the ancestral range.

Punctuated equilibrium explains the prevalence of stasis in the fossil record by linking macroevolution with speciation. This identification is necessary, according to Eldredge and Gould, because in a large interbreeding population something called "gene flow" hinders evolution. What this means is simply that the effect of favorable mutations is diluted by the sheer bulk of the population through which they must spread. This factor explains why species seem so unchanging in the fossil record: the population as a whole is *not* changing. The important evolutionary change occurs only among the peripheral isolates, who rejoin the stable ancestral population "suddenly" after forming a new species.

Most evolutionary biologists do not accept Eldredge and Gould's hypothesis that evolutionary change is closely associated with speciation. A great deal of variation can be obtained within a biological species (remember those dogs), whereas separate species are often very similar in visible characteristics. Speciation and change in form therefore seem to be different phenomena. Whether dilution or "gene flow" actually impedes change in large populations is the

---

[2] Terms like "rapidly" in this connection refer to geological time, and readers should bear in mind that 100,000 years is a brief period to a geologist. The punctuationalists' emphatic repudiation of "gradualism" is confusing, and tends to give the impression they are advocating saltationism. What they seem to mean is that the evolutionary change occurs over many generations by Darwin's step-by-step method, but in a relatively brief period of geological time. The ambiguity may be deliberate, however, for reasons that will be explained in this chapter.

subject of an apparently unresolvable theoretical dispute. Evidence that daughter populations form and then rejoin the parent species is lacking. According to Douglas Futuyma, "few if any" examples have been documented of an ancestral form persisting in the same region with a modified descendant.

For these and other reasons, orthodox neo-Darwinists prefer to explain sudden appearance on the traditional basis of gaps in the fossil record, and stasis as a reflection of "mosaic evolution" and "stabilizing selection." The former means that the soft body parts might have been evolving invisibly while the parts which fossilized stayed the same. The latter means that natural selection prevented change by eliminating all the innovations, sometimes for periods of millions of years and despite changing environmental conditions that ought to have encouraged adaptive innovation. Natural selection appears here in its formulation as a tautology with rather too much explanatory power, an invisible all-purpose explanation for whatever change or lack of change happened to occur.

If Darwinism enjoys the status of an *a priori* truth, then the problem presented by the fossil record is how Darwinist evolution always happened in such a manner as to escape detection. If, on the other hand, Darwinism is a scientific hypothesis which can be confirmed or falsified by fossil evidence, then the really important thing about the punctuationalism controversy is not the solution Gould, Eldredge, and Stanley proposed but the problem to which they drew attention. I see no reason to doubt that punctuationalism is a valid model for evolution in some cases. There are instances, such as the proliferation of fruitfly species in Hawaii, where it appears that rapid diversification has occurred following an initial migration of a parent species into a new region. The important question is not whether rapid speciation in peripheral isolates has occurred, however, but whether this mechanism can explain more than a relatively narrow range of modifications which cross the species boundary but do not involve major changes in bodily characteristics.

Consider the problem posed by Stanley's example of whales and bats, a mid-range case involving change within a single class. Nobody is proposing that an ancestral rodent (or whatever) became a whale or a bat in a single episode of speciation, with or without the aid of a mutation in its regulatory genes. Many intermediate species would have had to exist, some of which ought to have been numerous and long-lived. None of these appear in the fossil record. Of course the intermediates could have been very shortlived if they were not well fitted for survival, as would probably be the case with a creature midway in the process of changing legs to fins or wings. Raising this issue, however, adds nothing to the plausibility of the Darwinist scenario.

No doubt a certain amount of evolution could have occurred in such a way that it left no trace in the fossil record, but at some point we need more than ingenious excuses to fill the gaps. The discontinuities between the major groups—phyla, classes, orders—are not only pervasive, but in many cases immense. Was there never anything but invisible peripheral isolates in between?

The single greatest problem which the fossil record poses for Darwinism is the "Cambrian explosion" of around 600 million years ago. Nearly all the animal phyla appear in the rocks of this period, without a trace of the evolutionary ancestors that Darwinists require. As Richard Dawkins puts it, "It is as though they were just planted there, without any evolutionary history." In Darwin's time there was no evidence for the existence of pre-Cambrian life, and he conceded in *The Origin of Species* that "The case at present must remain inexplicable, and may be truly urged as a valid argument against the views here entertained." If his theory was true, Darwin wrote, the pre-Cambrian world must have "swarmed with living creatures."

In recent years evidence of bacteria and algae has been found in some of the earth's oldest rocks, and it is generally accepted today that these single-celled forms of life may have first appeared as long ago as four billion years. Bacteria and algae are "prokaryotes," which means each creature consists of a single cell without a nucleus and related organelles. More complex "eukaryote" cells (with a nucleus) appeared later, and then dozens of independent groups of multicellular animals appeared without any visible process of evolutionary development. Darwinist theory requires that there have been very lengthy sets of intermediate forms between unicellular organisms and animals like insects, worms, and clams. The evidence that these existed is missing, however, and with no good excuse.[3]

---

[3] The picture is clouded slightly by uncertainty over the status of the Ediacarans, a group of soft-bodied, shallow-water marine invertebrates found in rocks dating from shortly before the

The problem posed by the Cambrian explosion has become known to many contemporary readers due to the success of Gould's book *Wonderful Life*, describing the reclassification of the Cambrian fossils known as the Burgess Shale. According to Gould, the discoverer of the Burgess Shale fossils, Charles Walcott, was motivated to "shoehorn" them into previously known taxonomic categories because of his predisposition to support what is called the "artifact theory" of the pre-Cambrian fossil record. In Gould's words:

> Two different kinds of explanations for the absence of Precambrian ancestors have been debated for more than a century: the artifact theory (they did exist, but the fossil record hasn't preserved them), and the fast-transition theory (they really didn't exist, at least as complex invertebrates easily linked to their descendants, and the evolution of modern anatomical plans occurred with a rapidity that threatens our usual ideas about the stately pace of evolutionary change).

More recent investigation has shown that the Burgess Shale fossils include some 15 or 20 species that cannot be related to any known group and should probably be classified as separate phyla, as well as many other species that fit within an existing phylum but still manifest quite different body plans from anything known to exist later. The general picture of animal history is thus a burst of general body plans followed by extinction. No new phyla evolved thereafter. Many species exist today which are absent from the rocks of the remote past, but these all fit within general taxonomic categories present at the outset. The picture is one of evolution of a sort, but only within the confines of basic categories which themselves show no previous evolutionary history. Gould described the reclassification of the Burgess fossils as the "death knell of the artifact theory," because

---

Cambrian explosion. Some paleontologists have interpreted these as precursors to a few of the Cambrian groups. More recent studies by a paleontologist named Seilacher support the view, accepted by Gould, "that the Ediacaran fauna contains no ancestors for modern organisms, and that every Ediacaran animal shares a basic mode of organization quite distinct from the architecture of living groups." So interpreted, the Ediacarans actually demolish a standard Darwinist explanation for the absence of pre-Cambrian ancestors: that soft-bodied creatures would not fossilize. In fact many ancient soft-bodied fossils exist, in the Burgess Shale and elsewhere.

If evolution could produce ten new Cambrian phyla and then wipe them out just as quickly, then what about the surviving Cambrian groups? Why should they have had a long and honorable Precambrian pedigree? Why should they not have originated just before the Cambrian, as the fossil record, read literally, seems to indicate, and as the fast-transition theory proposes?

An orthodox Darwinist would answer that a direct leap from unicellular organisms to 25 to 50 complex animal phyla without a long succession of transitional intermediates is not the sort of thing for which a plausible genetic mechanism exists, to put it mildly. Gould is describing something he calls "evolution," but the picture is so different from what Darwin and his successors had in mind that perhaps a different term ought to be found. The Darwinian model of evolution is what Gould calls the "cone of increasing diversity." This means that the story of multicellular animal life should begin with a small number of species evolving from simpler forms. The dozens of different basic body plans manifested in the Cambrian fossils would then be the product of a long and gradual process of evolution from less differentiated beginnings. Nor should the cone have stopped expanding abruptly after the Cambrian explosion. If the disconfirming facts were not already known, any Darwinist would be confident that the hundreds of millions of years of post-Cambrian evolution would have produced many new phyla.

Instead we see the basic body plans all appearing first, many of these becoming extinct, and further diversification proceeding strictly within the boundaries of the original phyla. These original Cambrian groups have no visible evolutionary history, and the "artifact theory" which would supply such a history has to be discarded. Maybe a few evolutionary intermediates existed for some of the groups, although none have been conclusively identified, but otherwise just about all we have between complex multicellular animals and single cells is some words like "fast-transition." We can call this thoroughly un-Darwinian scenario "evolution," but we are just attaching a label to a mystery.

Sudden appearance and stasis of species in the fossil record is the opposite of what Darwinian theory would predict, and the pattern of extinctions is equally disappointing. There appear to have been a number of mass extinctions in the history of the earth, and debate still continues about what caused them. Two catastrophes in particular stand out: the Permian extinction of about 245 million years ago, which exterminated half the families of marine invertebrates and probably more than 90 per cent of all species; and the famous "K-T" extinction at the end of the Cretaceous era, about 65 million years ago, which exterminated the dinosaurs and a great deal else besides, including those ammonites whose disappearance Darwin conceded to have been wonderfully sudden.

According to Gould, paleontologists have known about these "great dyings" all along, but they have tried to minimize their importance because "our strong biases for gradual and continuous change force us to view mass extinctions as anomalous and threatening." Catastrophic explanations of extinction are making a strong comeback, however, and many researchers now report that the mass extinctions were more frequent, more rapid, and more profound in their effects than had previously been acknowledged.

Catastrophism is a controversial subject among geologists and paleontologists. Many scientific papers have argued that dinosaurs and ammonites were disappearing from the earth for millions of years before the meteorite impact which may have set off the K-T catastrophe. The stakes in this esoteric controversy are high, because Darwinism requires that old forms (the missing ancestors and intermediates) die out gradually as they are replaced by better adapted new forms. A record of extinction dominated by global catastrophes, in which the difference between survival and extinction may have been arbitrary, is as disappointing to Darwinist expectations as a record of sudden appearance followed by stasis.

There will be new controversies about the fossils before long, and probably anything written today will be outdated within a few years. The point to remember, however, is that the fossil problem for Darwinism is getting worse all the time. Darwinist paleontologists are indignant when creationists point this out, but what they write themselves is extraordinarily revealing. As usual, Gould is the most interesting commentator.

After attending a geological conference on mass extinctions, Gould wrote a remarkable essay reflecting on how the evidence was turning against Darwinism. He told his readers that he had long been puzzled by the lack of evidence of progressive development over time in the invertebrates with which he was most familiar. "We

can tell tales of improvement for some groups, but in honest moments we must admit that the history of complex life is more a story
of multifarious variation about a set of basic designs than a saga of
accumulating excellence." But Darwinist evolution should be a story
of improvement in fitness,[4] and so Gould regarded "the failure to
find a clear 'vector of progress' in life's history as the most puzzling
fact of the fossil record."

He thought the solution to the puzzle might lie in alternating
periods of evolution by punctuated equilibrium on the one hand,
and arbitrary extinction during catastrophes on the other. Under
these circumstances evolution would not be a story of gradual adaptive improvement, but rather "Evolutionary success must be as-
sessed among species themselves, not at the traditional Darwinian
level of struggling organisms within populations." Adopting without hesitation the "tautology" formulation of natural selection at the
species level, Gould proposed that "The reasons that species succeed are many and varied—high rates of speciation and strong
resistance to extinction, for example—and often involve no reference to traditional expectations for improvement in morphological
design."

Just about everyone who took a college biology course during the
last sixty years or so has been led to believe that the fossil record was
a bulwark of support for the classic Darwinian thesis, not a liability
that had to be explained away. And if we didn't take a biology class
we saw *Inherit the Wind* and laughed along with everybody else when
Clarence Darrow made a monkey out of William Jennings Bryan.
But I wonder if Bryan would have looked like such a fool if he could
have found a distinguished paleontologist having one of those "honest moments," and produced him as a surprise witness to tell the
jury and the theater audience that the fossil record shows a consistent pattern of sudden appearance followed by stasis, that life's
history is more a story of variation around a set of basic designs than

---

[4] Gould supported that point with a Darwin quote, but I will substitute a better one: "It may
be said that natural selection is daily and hourly scrutinising, throughout the world, every
variation, even the slightest; rejecting that which is bad, preserving and adding up all that is
good: silently and insensibly working, whenever and wherever opportunity offers, at the
improvement of each organic being in relation to its organic and inorganic condition of life."
In later editions, Darwin added the word "metaphorically" to the sentence, apparently
realizing that he had written of natural selection as if it were an intelligent, creative being.

one of accumulating improvement, that extinction has been predominantly by catastrophe rather than gradual obsolescence, and that orthodox interpretations of the fossil record often owe more to Darwinist preconception than to the evidence itself. Imagine the confusion that Bryan could have caused by demanding the right to read his own preferred evidence into those famous gaps! Why not, if Darwin could do it?

Paleontologists seem to have thought it their duty to protect the rest of us from the erroneous conclusions we might have drawn if we had known the actual state of the evidence. Gould described "the extreme rarity of transitional forms in the fossil record" as "the trade secret of paleontology." Steven Stanley explained that the doubts of paleontologists about gradualistic evolution were for long years "suppressed." He wrote that the process began with T. H. Huxley himself, who muted "his negative attitudes toward gradual change and natural selection," presumably because "as a believer, Huxley was not inclined to aid those who were disposed to throw the baby of evolution out with the bathwater of gradualistic natural selection." But why would Huxley fear that, unless the baby and the bathwater were impossible to separate?

Niles Eldredge has been even more revealing: "We paleontologists have said that the history of life supports [the story of gradual adaptive change], all the while really knowing that it does not." But how could a deception of this magnitude possibly have been perpetrated by the whole body of a respected science, dedicated almost by definition to the pursuit of truth? Eldredge's explanation is all too believable to anyone who is familiar with the ways of the academic world:

Each new generation, it seems, produces a few young paleontologists eager to document examples of evolutionary change in their fossils. The changes they have always looked for have, of course, been of the gradual, progressive sort. More often than not their efforts have gone unrewarded—their fossils, rather than exhibiting the expected pattern, just seem to persist virtually unchanged. . . . This extraordinary conservatism looked, to the paleontologist keen on finding evolutionary change, as if no evolution had occurred. Thus studies documenting conservative persistence rather than gradual evolutionary change were considered failures, and, more often than not, were not even published. Most paleontologists were aware of the stability, the lack of change we call stasis. . . . But insofar as evolution itself is concerned, paleontologists usually saw stasis as "no results" rather than as a contradiction of the prediction of gradual, progressive evolutionary change. Gaps in the record continue (to this day) to be invoked as the prime reason why so few cases of gradual change are found.

Gould wrote in the same vein that "When Niles Eldredge and I proposed the theory of punctuated equilibrium in evolution, we did so to grant stasis in phylogenetic lineages the status of 'worth reporting'—for stasis had previously been ignored as nonevidence of evolution, though all paleontologists knew its high relative frequency." What Gould and Eldredge had to avoid, however, was what Eldredge described as "the not-unreasonable relegation to the lunatic fringe that some paleontologists in the past had suffered when they too saw something out of kilter between contemporary evolutionary theory, on the one hand, and patterns of change in the fossil record on the other." In short, they had to avoid seeming to embrace saltationism.

In the preceding chapter I mentioned the paleontologist Otto Schindewolf, whose saltationism extended to the extreme of proposing that the first bird must have hatched from a reptile's egg. George Gaylord Simpson reviewed Schindewolf's book disapprovingly, but he conceded that its author's bizarre conclusions were based upon a thorough knowledge of the fossil evidence. The trouble with Schindewolf was that he made no attempt to impose an interpretation upon the fossil evidence which could be accepted by the geneticists, or perhaps he relied too much upon the approval of the geneticist Richard Goldschmidt. He just went ahead and published what the fossils told him, and the fossils said "saltation."

Paleontologists who have to work under the influence of neo-Darwinism do not have the same freedom to draw whatever conclusions their evidence leads them to. Eldredge has described the paleontologist's dilemma frankly: "either you stick to conventional theory despite the rather poor fit of the fossils, or you focus on the empirics and say that saltation looks like a reasonable model of the evolutionary process—in which case you must embrace a set of rather dubious biological propositions." Paleontology, it seems, is a discipline in which it is sometimes unseemly to "focus on the empirics." On the other hand, one can't just go out and manufacture evidence of Darwinist evolution, and Eldredge wrote movingly about how this combination of restrictions makes it difficult to pursue a successful career:

> Complicating the normal routine is the hassle of obtaining a Ph.D. A piece of doctoral research is really an apprenticeship, and the dissertation a comprehensive report that shows the candidate's ability to frame, and successfully pursue, an original piece of scientific research. Sounds reasonable, but the pressure for results, positive results, is enormous.

In these frustrating circumstances, paleontologists clearly needed to find a theory that would allow them to report their projects as successful, but they felt constrained to operate within the boundaries of the neo-Darwinian synthesis. What was required was a theory that was saltationist enough to allow the paleontologists to publish, but gradualistic enough to appease the Darwinists. Punctuated equilibrium accomplishes this feat of statesmanship by making the process of change inherently invisible. You can imagine those peripheral isolates changing as much and as fast as you like, because no one will ever see them.

Gould and Eldredge have consistently described punctuated equilibrium as a Darwinist theory, not a saltationist repudiation of Darwinism. On the other hand, it is easy to see how some people got the impression that saltationism was at least being hinted, if not explicitly advocated. Gould and Eldredge put two quotes by T. H. Huxley on the front of their 1977 paper, both complaints about Darwin's refusal to allow a little "saltus" in his theory. At about the same time, Gould independently endorsed a qualified saltationism and predicted Goldschmidt's vindication.

The trouble with saltationism, however, is that when closely examined it turns out to be only a meaningless middle ground somewhere between evolution and special creation. As Richard Dawkins put it, you can call the Biblical creation of man from the dust of the earth a saltation. In terms of fossil evidence, saltation just means that a new form appeared out of nowhere and we haven't the faintest idea how. As a scientific theory, "saltationist evolution" is just what Darwin called it in the first place: rubbish. Gould and Eldredge understand that, and so despite hints of saltationism (particularly by Gould) they have always kept open their lines of retreat to orthodox Darwinian gradualism.

This raises the most basic question of all. If there are so many problems with Darwinism, and no satisfactory alternative within the framework of evolution, why not reevaluate the framework? What makes our scientists so absolutely certain that everything really *did* evolve from simple beginnings?

Chapter Five

The Fact of Evolution

DARWINISTS CONSIDER EVOLUTION to be a fact, not just a theory, because it provides a satisfying explanation for the pattern of relationship linking all living creatures—a pattern so identified in their minds with what they consider to be the necessary *cause* of the pattern—descent with modification—that, to them, biological relationship *means* evolutionary relationship.

Biological classification is about as controversial a subject as religion or politics, but some basic principles are generally accepted. Biologists classify animals (and other organisms) by taxonomic categories such as families, orders, classes, and phyla. A superficial classification might group the whale, the penguin, and the shark together as aquatic creatures, and birds, bats, and bees together as flying creatures. But the basic body design of birds, bats, and bees is fundamentally different, their reproductive systems are different, and even their wings are similar only in the sense that they are all fit for flying. Accordingly, all taxonomists agree that the bat and the

63

whale should be grouped with the horse and the monkey as mammals, despite the enormous differences in behavior and adaptive mechanisms. Bees are built on a fundamentally different body plan from vertebrates of any kind, and go into a different series of groupings altogether.

Biologists before and after Darwin have generally sensed that in classifying they were not merely forcing creatures into arbitrary categories, but discovering relationships that are in some sense real. Some pre-Darwinian taxonomists expressed this sense by saying that whales and bats are superficially like fish and birds but they are *essentially* mammals—that is, they conform in their "essence" to the mammalian "type." Similarly, all birds are essentially birds, whether they fly, swim, or run. The principle can be extended up or down the scale of classification: St. Bernards and dachshunds are essentially dogs, despite the visible dissimilarity, and sparrows and elephants are essentially vertebrates.

Essentialism did not attempt to explain the cause of natural relationships, but merely described the pattern in the language of Platonic philosophy. The essentialists knew about fossils and hence were aware that different kinds of creatures had lived at different times. The concept of evolution did not make sense to them, however, because it required the existence of numerous intermediates—impossible creatures that were somewhere in transition from one essential state to another. Essentialists therefore attributed the common features linking each class not to inheritance from common ancestors, but to a sort of blueprint called the "Archetype," which existed only in some metaphysical realm such as the mind of God.

Darwin proposed a naturalistic explanation for the essentialist features of the living world that was so stunning in its logical appeal that it conquered the scientific world even while doubts remained about some important parts of his theory. He theorized that the discontinuous groups of the living world were the descendants of long-extinct common ancestors. Relatively closely related groups (like reptiles, birds, and mammals) shared a relatively recent common ancestor; all vertebrates shared a more ancient common ancestor; and all animals shared a still more ancient common ancestor. He then proposed that the ancestors must have been linked to their descendants by long chains of transitional intermediates, also extinct. According to Darwin:

We may thus [by extinction] account even for the distinctness of whole classes from each other—for instance, of birds from all other vertebrate animals—by the belief that many ancient forms of life have been utterly lost, through which the early progenitors of birds were formerly connected with the early progenitors of the other vertebrate classes.

This theory of descent with modification made sense out of the pattern of natural relationships in a way that was acceptable to philosophical materialists. It explained why the groups seemed to be part of the natural framework rather than a mere human invention—to the Darwinist imagination, they are literally families. When combined with the theory of natural selection, it explained the difference between the common features that are relevant to classification (*homologies*) and those that are not (*analogies*). The former were relics of the common ancestor; the latter evolved independently by natural selection to provide very different creatures with superficially similar body parts that were useful to such adaptive strategies as flight and swimming. In Darwin's historic words:

All the . . . difficulties in classification are explained . . . on the view that the natural system is founded on descent with modification: that the characters which naturalists consider as showing true affinity between any two or more species, are those which have been inherited from a common parent, and in so far, all true classification is genealogical; that community of descent is the hidden bond which naturalists have been unconsciously seeking, and not some unknown plan of creation, or the enunciation of general propositions, and the mere putting together and separating objects more or less alike.

Darwin ended his chapter by saying that the argument from classification was so decisive that on that basis alone he would adopt his theory even if it were unsupported by other arguments. That confidence explains why Darwin was undiscouraged by the manifold difficulties of the fossil record: his logic told him that descent with modification had to be the explanation for the "difficulties in classification," regardless of any gaps in the evidence. The same logic inspires today's Darwinists, when they shrug off critics who claim that one element or another in the theory is doubtful. "Say what you will against every detail," they respond, "still, nothing in biology makes sense except in the light of evolution."

Darwin's theory unquestionably has impressive explanatory power, but how are we to tell if it is *true*? If we define "evolution" simply as "whatever produces classification," then evolution is a fact in the same sense that classification is a fact. This is another tautology, however, and as such it has no genuine explanatory value. In this form the theory is supported mainly by the semantic implications of the word "relationship." Darwinists assume that the relationship between, say, bats and whales is similar to that between siblings and cousins in human families. Possibly it is, but the proposition is not self-evident.

Descent with modification could be something much more substantial than a tautology or a semantic trick. It could be a testable scientific hypothesis. If common ancestors and chains of linking intermediates once existed, fossil studies should be able, at least in some cases, to identify them. If it is possible for a single ancestral species to change by natural processes into such different forms as a shark, a frog, a snake, a penguin, and a monkey, then laboratory science should be able to discover the mechanism of change.

If laboratory science cannot establish a mechanism, and if fossil studies cannot find the common ancestors and transitional links, then Darwinism fails as an empirical theory. But Darwinists suppress consideration of that possibility by invoking a distinction between the "fact" of evolution and Darwin's particular theory. Objections based upon the fossil record and the inadequacy of the Darwinist mechanism go only to the theory, they argue. Evolution itself (the logical explanation for relationships) remains a fact, by which they seem to mean it is an inescapable deduction from the fact of relationship. Stephen Jay Gould's influential article, "Evolution as Fact and Theory" explains the distinction by citing the fact and theory of gravity:

> Facts are the world's data. Theories are structures of ideas that explain and interpret facts. Facts do not go away while scientists debate rival theories for explaining them. Einstein's theory of gravitation replaced Newton's, but apples did not suspend themselves in mid-air pending the outcome. And human beings evolved from ape-like

ancestors whether they did so by Darwin's proposed mechanism or by some other, yet to be identified.

The analogy is spurious. We observe directly that apples fall when dropped, but we do not observe a common ancestor for modern apes and humans. What we *do* observe is that apes and humans are physically and biochemically more like each other than they are like rabbits, snakes, or trees. The ape-like common ancestor is a hypothesis in a *theory*, which purports to explain how these greater and lesser similarities came about. The theory is plausible, especially to a philosophical materialist, but it may nonetheless be false. The true explanation for natural relationships may be something much more mysterious.

Because Gould draws the line between fact and theory in the wrong place, the distinction is virtually meaningless. The theory to him is merely the theory of natural selection, and the "fact" is the fact that evolution may occur by chance mechanisms without influence from selection. Gould explains the distinction by observing that

> while no biologist questions the importance of natural selection, many now doubt its ubiquity. In particular, many evolutionists argue that substantial amounts of genetic change may not be subject to natural selection and may spread through populations at random.

As Gould acknowledges, however, Darwin always insisted that natural selection was only *one* of the mechanisms of evolution, and complained bitterly when he was accused of writing that selection is ubiquitous. The "fact" that Gould describes is therefore nothing but Darwin's *theory* rightly understood: evolution is descent with modification propelled by random genetic changes, with natural selection providing whatever guidance is needed to produce complex adaptive structures like wings and eyes.[1] The creative power of

---

[1] Readers should not be misled by the daring speculations of a few paleontologists like Gould and Steven Stanley, who flirt with macromutational alternatives to Darwinist gradualism. No genuine alternative to Darwinism is in prospect. From T. H. Huxley's time to the present, there have been paleontologists who acknowledged that the fossil record is inconsistent with strict Darwinism. To mitigate the difficulty, they have tried to describe a saltationist alternative in language the purists could tolerate.

The fossil problem, however, is not the main issue. A fact or theory of evolution would not

natural selection is then assured because it is a necessary implication of the "fact" that evolution has produced all the wonders of biology. Recasting the theory as fact serves no purpose other than to protect it from falsification.

Nobody needs to prove that apples fall down rather than up, but Gould provides three proofs for the "fact of evolution." The first proof is microevolution:

> First, we have abundant, direct, observational evidence of evolution in action, from both field and laboratory. This evidence ranges from countless experiments on change in nearly everything about fruit flies subjected to artificial selection in the laboratory to the famous populations of British moths that became black when industrial soot darkened the trees upon which the moths rest. (Moths gain protection from sharp-sighted bird predators by blending into the background.) Creationists do not deny these observations: how could they? Creationists have tightened their act. They now argue that God only created "basic kinds," and allowed for limited evolutionary meandering within them. Thus toy poodles and Great Danes come from the dog kind and moths can change color, but nature cannot convert a dog to a cat or a monkey to a man.

Gould is right: everyone agrees that microevolution occurs, including creationists. Even creation-scientists concur, not because they "have tightened their act," but because their doctrine has always been that God created basic kinds, or types, which subsequently diversified. The most famous example of creationist microevolution involves the descendants of Adam and Eve, who have diversified from a common ancestral pair to create all the diverse races of the human species.

The point in dispute is not whether microevolution happens, but whether it tells us anything important about the processes responsible for creating birds, insects, and trees in the first place. Gould himself has written that even the first step toward macroevolution (speciation) requires more than the accumulation of micromuta-

---

be worth much if it could not explain the origin of complex biological structures, and nobody has found a naturalistic alternative to micromutation and selection for that purpose. Even Gould has to rely upon orthodox Darwinism when he looks away from the fossil problem and turns to justifying "evolution" as a general explanation for the origin of complex biological structures like wings and eyes.

tions. Instead of explaining how the peppered moth variations bear on the kind of evolution that really matters, however, he changes the subject and takes a swipe at creationists.[2]

Other Darwinists who do not simply ignore the problem resort to bad philosophy to evade it. For example, Mark Ridley asserts that "All that is needed to prove evolution is observed microevolution added to the philosophical doctrine of uniformitarianism which (in the form that is needed here) underlies all science."

But what sort of proof is this? If our philosophy demands that small changes add up to big ones, then the scientific evidence is irrelevant. Scientists like to assume that the laws of nature were always and everywhere uniform, because otherwise they could not make inferences about what happened in the distant past or at the opposite end of the universe. They do not assume that the rules which govern activity at one level of magnitude necessarily apply at all other levels. The differences between Newtonian physics, relativity, and quantum mechanics show how unjustified such an assumption would be. What the Darwinists need to supply is not an arbitrary philosophical principle, but a scientific theory of how macroevolution can occur.

Much confusion results from the fact that a single term—"evolution"—is used to designate processes that may have little or nothing in common. A shift in the relative numbers of dark and light moths in a population is called evolution, and so is the creative process that produced the cell, the multicellular organism, the eye, and the human mind. The semantic implication is that evolution is fundamentally a single process, and Darwinists enthusiastically exploit that implication as a substitute for scientific evidence. Even the separation of evolution into its "micro" and "macro" varieties—which Darwinists generally resist—implies that all the creative processes involved in life comprise a single, two-part phenomenon that will be adequately understood when we discover a process that makes new species from existing ones. Possibly this is the case, but more probably it is not. The vocabulary of Darwinism inherently

---

[2] Creationist-bashing as a substitute for evidence is common in Darwinist polemics. For example, Isaac Asimov's 884-page *New Guide to Science* has a half-page section on the evidence for Darwinism, which cites the peppered moth example as sufficient to prove the whole theory. This is preceded by almost three pages abusing creationists. The lapse from professionalism is striking, because on other topics the book is admirably scientific.

limits our comprehension of the difficulties by misleadingly covering them with the blanket term "evolution."

Gould's second argument, and the centerpiece of his case for the
"fact" of evolution, is the argument from imperfection:

> The second argument—that the imperfection of nature reveals
> evolution—strikes many people as ironic, for they feel that evolution
> should be most elegantly displayed in the nearly perfect adaptation
> expressed by some organisms—the camber of a gull's wing, or but-
> terflies that cannot be seen in ground litter because they mimic leaves
> so precisely. But perfection could be imposed by a wise creator or
> evolved by natural selection. Perfection covers the tracks of past his-
> tory. And past history—the evidence of descent—is the mark of
> evolution.
>
> Evolution lies exposed in the imperfections that record a history of
> descent. Why should a rat run, a bat fly, a porpoise swim, and I type
> this essay with structures built of the same bones unless we all inher-
> ited them from a common ancestor? An engineer, starting from
> scratch, could design better limbs in each case. Why should all the
> large native mammals of Australia be marsupials, unless they de-
> scended from a common ancestor on this island continent? Marsu-
> pials are not "better," or ideally suited for Australia; many have been
> wiped out by placental animals imported by man from other conti-
> nents. . . .

Gould here merely repeats Darwin's explanation for the existence of natural groups—the theory for which we are seeking
confirmation—and gives it a theological twist. A proper Creator
should have designed each kind of organism from scratch to achieve
maximum efficiency. This speculation is no substitute for scientific
evidence establishing the reality of the common ancestors. It also
does nothing to confirm the natural process by which the transformation from ancestral to descendant forms supposedly occurred. It
is Darwin, after all, who banished speculation about the "unknown
plan of creation" from science.

Douglas Futuyma also leans heavily on the "God wouldn't have
done it" theme, citing examples from vertebrate embryology:

> Why should species that ultimately develop adaptations for utterly
> different ways of life be nearly indistinguishable in their early stages?

How does God's plan for humans and sharks require them to have almost identical embryos? Why should terrestrial salamanders, if they were not descended from aquatic ancestors, go through a larval stage entirely within the egg, with gills and fins that are never used, and then lose these features before they hatch?

These are rhetorical questions, but they point to legitimate starting points for investigation. The features Futuyma cites may exist because a Creator employed them for some inscrutable purpose; or they may reflect inheritance from specific common ancestors; or they may be due to some as yet unimagined process which science may discover in the future. The task of science is not to speculate about why God might have done things this way, but to see if a material cause can be established by empirical investigation. If evolutionary biology is to be a science rather than a branch of philosophy, its theorists have to be willing to ask the scientific question: *How can Darwin's hypothesis of descent with modification be confirmed or falsified?*

Most of the evidence relied upon by today's Darwinists was known to Darwin's great contemporary, the Swiss-born Harvard scientist Louis Agassiz. Agassiz's resistance to Darwinism did not stem from any failure to understand the evidence that made the theory so beguiling for others. Writing not long after the publication of *The Origin of Species,* he concluded that

it is evident that there is a manifest progress in the succession of beings on the surface of the earth. This progress consists in an increasing similarity to the living fauna, and among the Vertebrates, especially, in their increasing resemblance to man.

But this connection is not the consequence of a direct lineage between the faunas of different ages. There is nothing like parental descent connecting them. The Fishes of the Paleozoic Age are in no respect the ancestors of the Reptiles of the Secondary Age, nor does Man descend from the Mammals which preceded him in the Tertiary Age. The link by which they are connected is of a higher and immaterial nature; and their connection is to be sought in the view of the Creator himself, whose aim ... was to introduce Man upon the surface of our globe.

Agassiz's theological opinion is no more relevant to the empirical question than Gould's, but we may put it aside without affecting the strictly scientific content of his conclusion. His empirical point was that whatever might have caused the appearance of progression in the vertebrate sequence, the evidence of the fossil record is that it was *not* descent with modification.

That brings us to Gould's third proof, which takes us back to the fossil record. Gould concedes that fossil evidence of macroevolutionary transformations has rarely been found, but he insists that there are at least two instances in the vertebrate sequence where such transformations can be confirmed. One example is the "mammal-like reptiles," which, as the name implies, appear to be intermediates in the reptile-to-mammal transformation. The other is the hominids, or "ape-men," which are accepted by mainstream science as genuine predecessors of modern humans. This fossil evidence is the subject of the next chapter.

Chapter Six

The Vertebrate Sequence

DARWINISTS CLAIM THAT amphibians and modern fish descended from an ancestral fish; that reptiles descended from an amphibian ancestor; and that birds and mammals descended separately from reptile ancestors. Finally, they say that humans and modern apes had a common simian ancestor, from which modern humans descended through transitional intermediates that have been positively identified. According to Gould, fossils in the reptile-to-mammal and ape-to-human transitions provide decisive confirmation of the "fact of evolution."

Before going to the evidence I have to impose an important condition which is sure to make Darwinists very uncomfortable. It is that the evidence must be evaluated independently of any assumption about the truth of the theory being tested.

Paleontology, as we saw in Chapter Four, has taken Darwinian descent as a deductive certainty and has sought to flesh it out in detail rather than to test it. Success for fossil experts who study evolution has meant success in identifying ancestors, which provides an incentive for establishing criteria that will permit ancestors to be identified. Gareth Nelson of the American Museum of Natural History has expressed in plain language what this has meant in practice:

> "We've got to have some ancestors. We'll pick those." Why? "Because we know they have to be there, and these are the best candidates." That's by and large the way it has worked. I am not exaggerating.

Obviously, "ancestors" cannot confirm the theory if they were labelled as such only because the theory told the researchers that ancestors had to be there.

Now let's look at the vertebrate sequence.

Fish to Amphibians

The story to be tested is that a fish species developed the ability to climb out of the water and move on land, while evolving the peculiar reproductive system of amphibians and other amphibian features more or less concurrently. No specific fossil fish species has been identified as an amphibian ancestor, but there is an extinct order of fish known as the rhipidistians which Darwinists frequently describe as an "ancestral group." The rhipidistians have skeletal features resembling those of early amphibians, including bones that look like they could have evolved into legs. But according to Barbara J. Stahl's comprehensive textbook, *Vertebrate History*, "none of the known fishes is thought to be directly ancestral to the earliest land vertebrates. Most of them lived after the first amphibians appeared, and those that came before show no evidence of developing the stout limbs and ribs that characterized the primitive tetrapods."

In 1938, a coelacanth (pronounced see-la-kanth), an ancient fish thought to have been extinct for about seventy million years, was caught by fishermen in the Indian Ocean. Many paleontologists considered the coelacanth to be closely related to the rhipidistians, and thus a living specimen was expected to shed light on the soft body parts of the immediate ancestors of amphibians. When the modern coelacanth was dissected, however, its internal organs showed no signs of being preadapted for a land environment and gave no indication of how it might be possible for a fish to become an amphibian. The experience suggests that a rhipidistian fish might be equally disappointing to Darwinists if its soft body parts could be examined.

Amphibians to Reptiles

No satisfactory candidates exist to document this transition. There are fossil amphibians called *Seymouria* that have some reptile-like skeletal characteristics, but they appear too late in the fossil record and recent evidence indicates that they were true amphibians. The transition is in any case one which would be hard to confirm with fossils, because the most important difference between amphibians and reptiles involves the unfossilized soft parts of their reproductive systems. Amphibians lay their eggs in water and the larvae undergo a complex metamorphosis before reaching the adult stage. Reptiles lay a hard shell-cased egg and the young are perfect replicas of adults on first emerging. No explanation exists for how an amphibian could have developed a reptilian mode of reproduction by Darwinian descent.

Reptiles to Mammals

We come at last to the crown jewel of the fossil evidence for Darwinism, the famous mammal-like reptiles cited by Gould and many others as conclusive proof. The large order *Therapsida* contains many fossil species with skeletal features that appear to be intermediate between those of reptiles and mammals. At the boundary, fossil reptiles and mammals are difficult to tell apart. The usual criterion is that a fossil is considered reptile if its jaw contains several bones, of which one, the articular bone, connects to the quadrate bone of the skull. If the lower jaw consists of a single dentary bone, connecting to the squamosal bone of the skull, the fossil is classified as a mammal.

In this critical feature of jaw structure, and in some other features, various "therapsids" approach the mammalian characteristics so closely that in some cases they could be reasonably classified as either reptiles or mammals. Gould's vivid description brings out the importance of this:

The lower jaw of reptiles contains several bones, that of mammals only one. The non-mammalian jawbones are reduced, step by step, in mammalian ancestors until they become tiny nubbins located at the back of the jaw. The 'hammer' and 'anvil' bones of the mammalian ear are descendants of these nubbins. How could such a transition be accomplished? the creationists ask. Surely a bone is either entirely in the jaw or in the ear. Yet paleontologists have discovered two transitional lineages of therapsids (the so-called mammal-like reptiles) with a double jaw joint—one composed of the old quadrate and articular bones (soon to become the hammer and anvil), the other of the squamosal and dentary bones (as in modern mammals).

We may concede Gould's narrow point, but his more general claim that the mammal-reptile transition is thereby established is another matter. Creatures have existed with a skull bone structure intermediate between that of reptiles and mammals, and so the transition with respect to this feature is possible. On the other hand, there are many important features by which mammals differ from reptiles besides the jaw and ear bones, including the all-important reproductive systems. As we saw in other examples, convergence in skeletal features between two groups does not necessarily signal an evolutionary transition.

Douglas Futuyma makes a confident statement about the therapsids that actually reveals how ambiguous the therapsid fossils really are. He writes that "The gradual transition from therapsid reptiles to mammals is so abundantly documented by scores of species in every stage of transition that it is impossible to tell which therapsid species were the actual ancestors of modern mammals." But large numbers of eligible candidates are a plus only to the extent that they can be placed in a single line of descent that could conceivably lead from a particular reptile species to a particular early mammal descendant. The presence of similarities in many different species that are outside of any possible ancestral line only draws attention to the fact that skeletal similarities do not necessarily imply ancestry. The notion that mammals-in-general evolved from reptiles-in-general through a broad clump of diverse therapsid lines is not Darwinism. Darwinian transformation requires a single line of ancestral descent.

It seems that the mammal-like qualities of the therapsids were distributed widely throughout the order, in many different subgroups which are mutually exclusive as candidates for mammal ancestors. An artificial line of descent can be constructed, but only by arbitrarily mixing specimens from different subgroups, and by arranging them out of their actual chronological sequence. If our hypothesis is that mammals evolved from therapsids only once (a point to which I shall return), then most of the therapsids with mammal-like characteristics were not part of a macroevolutionary transition. If most were not then perhaps all were not.

The case for therapsids as an ancestral chain linking reptiles to mammals would be a great deal more persuasive if the chain could be attached to something specific at either end. Unfortunately, important structural differences among the early mammals make it just as difficult to pick a specific mammal descendant as it is to pick any specific therapsid ancestors. This baffling situation led some paleontologists to consider a disturbing theory that mammals, long assumed to be a natural "monophyletic" group (that is, descended from a common mammalian ancestor) were actually several groups which had evolved separately from different lines of therapsids.

Turning mammals into a polyphyletic group would make therapsids more plausible as ancestors, but only at the unacceptable cost of undermining the Darwinist argument that mammalian homologies are relics of common ancestry. Whether mammals evolved more than once remains an open question as far as fossils are concerned, but the prestigious George Gaylord Simpson lowered the stakes considerably by deciding that a group could reasonably be considered monophyletic if it descended from a single unit of lower rank in the taxonomic hierarchy. Having arisen from the order *Therapsida*, the class *Mammalia* was acceptable as a natural unit.

If one does not stop with the reptile-mammal transition but continues the attempt to provide a coherent account of macroevolution into the mammal class itself, it becomes immediately apparent that there is a great deal more to explain than the differences in jaw and ear bone structure between reptiles and mammals. The mammal class includes such diverse groups as whales, porpoises, seals, polar bears, bats, cattle, monkeys, cats, pigs, and opossums. If mammals are a monophyletic group, then the Darwinian model requires that every one of the groups have descended from a single unidentified small land mammal. Huge numbers of intermediate species in the direct line of transition would have had to exist, but the fossil record fails to record them.

Reptile to Bird

Archaeopteryx ("old wing"), a fossil bird which appears in rocks estimated to be 145 million years old, was discovered soon after the publication of *The Origin of Species*, and it thus helped enormously to establish the credibility of Darwinism and to discredit skeptics like Agassiz. *Archaeopteryx* has a number of skeletal features which suggest a close kinship to a small dinosaur called *Compsognathus*. It is on the whole bird-like, with wings, feathers, and wishbone, but it has claws on its wings and teeth in its mouth. No modern bird has teeth, although some ancient ones did, and there is a modern bird, the hoatzin, which has claws.

Archaeopteryx is an impressive mosaic. The question is whether it is proof of a reptile (dinosaur) to bird transition, or whether it is just one of those odd variants, like the contemporary duck-billed platypus, that have features resembling those of another class but are not transitional intermediates in the Darwinian sense. Until very recently, the trend among paleontologists was to regard *Archaeopteryx* as an evolutionary dead end rather than as the direct ancestor of modern birds. The next oldest bird fossils were specialized aquatic divers that did not look like they could be its direct descendants.[1]

The picture has changed somewhat following discoveries of fossil birds, one in Spain and the other in China, in rocks dated at 125 million and 135 million years. The new specimens have reptilian skeletal features which qualify them as possible intermediates between *Archaeopteryx* and certain modern birds. The evidence, however, is too fragmentary to justify any definite conclusions. According to a 1990 review article by Peter Wellnhofer, a recognized authority, it is impossible to determine whether *Archaeopteryx* actually was the ancestor of modern birds. Wellnhofer concludes that "this correlation is not of major importance," because the *Ar-*

---

[1] A paleontologist named Chatterjee claims to have found fossil evidence of a bird he calls Protoavis, in Texas rocks estimated to be 225 million years old. Bird fossils substantially older than 145 million years would disqualify *Archaeopteryx* as a bird ancestor, but Chatterjee's claim has been disputed.

chaeopteryx specimens "provide clues as to how birds evolved," and because "They are documents without which the idea of evolution would not be as powerful."

In *Archaeopteryx* we therefore have a possible bird ancestor rather than a certain one. As in the cases of mammals, there is plenty of difficulty in imagining how any single ancestor could have produced descendants as varied as the penguin, the hummingbird, and the ostrich, through viable intermediate stages. The absence of fossil evidence for the transitions is more easily excused, however, because birds pursue a way of life that ensures that their bodies will rarely be fossilized.

Archaeopteryx is on the whole a point for the Darwinists, but how important is it? Persons who come to the fossil evidence as convinced Darwinists will see a stunning confirmation, but skeptics will see only a lonely exception to a consistent pattern of fossil disconfirmation. If we are testing Darwinism rather than merely looking for a confirming example or two, then a single good candidate for ancestor status is not enough to save a theory that posits a worldwide history of continual evolutionary transformation.

Whatever one concludes about *Archaeopteryx*, the origin of birds presents many mysteries. Flight had to evolve, along with the intricate feathers and other specialized equipment, including the distinctive avian lung. Possibly birds did somehow develop from dinosaur predecessors, with *Archaeopteryx* as a way station, but even on this assumption we do not know what mechanism could have produced all the complex and interrelated changes that were necessary for the transformation.

From Apes to Humans

In the 1981 "Fact and Theory" article discussed in the preceding chapter, Gould cited the "half-dozen human species discovered in ancient rocks" as proof that humans evolved from apes. When he published a revised version of the same argument in 1987, the number of species had been reduced to five, one of which was Homo sapiens itself, but the point was the same:

> Would God—for some inscrutable reason, or merely to test our faith—create five species, one after the other (*Australopithecus af-*

arensis, A. africanus, Homo habilis, H. Erectus, and *H. Sapiens*), to mimic a continuous trend of evolutionary change?[2]

That way of putting the question makes it sound as if Darwin proposed his theory because the presence of an abundance of fossil intermediates between apes and humans required some explanatory hypothesis. Of course what actually happened is that the theory was accepted first, and the supporting evidence was discovered and interpreted in the course of a determined effort to find the "missing links" that the theory demanded. The question this sequence of events raises is not whether God has been planting fossil evidence to test our faith in Genesis, but whether the Darwinist imagination might have played an important role in construing the evidence which has been offered to support Darwin's theory.

Physical anthropology—the study of human origins—is a field that throughout its history has been more heavily influenced by subjective factors than almost any other branch of respectable science. From Darwin's time to the present the "descent of man" has been a cultural certainty begging for empirical confirmation, and worldwide fame has been the reward for anyone who could present plausible fossil evidence for missing links. The pressure to find confirmation was so great that it led to one spectacular fraud, Piltdown man—which British Museum officials zealously protected from unfriendly inspection, allowing it to perform forty years of useful service in molding public opinion.

Museum reconstructions based on the scanty fossil evidence have had a powerful impact on the public imagination, and the fossils themselves have had a similar effect upon the anthropologists. The psychological atmosphere that surrounds the viewing of hominid fossils is uncannily reminiscent of the veneration of relics at a medi-

---

[2] The four ape-man species that Gould cites include the two *Australopithecines* on the ape side of the boundary, which had ape brains but are supposed to have walked upright, and the larger-brained *Homo* specimens. Louis Leakey's *Homo habilis* (handy man) is at the borderline and was granted *Homo* status mainly because it was found at a site with primitive tools, which it is presumed to have used. Readers who learned about this subject in school may be surprised to find out that *Neanderthal* man is frequently considered a subgroup within our own species and *Cro-Magnon* man is simply modern man. Some other familiar names were either dropped from the pantheon or absorbed into the four species. Hominid fossil classification is a fiercely controversial subject and was in chaos until the ubiquitous Ernst Mayr stepped in and set the ground rules.

eval shrine. That is just how Roger Lewin described the scene at the 1984 *Ancestors* exhibition at the American Museum of Natural History, an unprecedented showing of original fossils relating to human evolution from all over the world.

The "priceless and fragile relics" were carried by anxious curators in first-class airplane seats and brought to the Museum in a VIP motorcade of limousines with police escort. Inside the Museum, the relics were placed behind bullet-proof glass to be admired by a select preview audience of anthropologists, who spoke in hushed voices because "It was like discussing theology in a cathedral." A sociologist observing this ritual of the anthropologist tribe remarked, "Sounds like ancestor worship to me."

Lewin considers it understandable that anthropologists observing the bones of their ancestors should be more emotionally involved with their subject than other kinds of scientists. "There *is* a difference. There *is* something inexpressibly moving about cradling in one's hands a cranium drawn from one's own ancestry." Lewin is absolutely correct, and I can't think of anything more likely to detract from the objectivity of one's judgment. Descriptions of fossils from people who yearn to cradle their ancestors in their hands ought to be scrutinized as carefully as a letter of recommendation from a job applicant's mother. In his book *Human Evolution*, Lewin reports numerous examples of the subjectivity that is characteristic of human origins research, leading him to conclude that the field is invisibly but constantly influenced by humanity's shifting self-image. In plain English, that means that we see what we expect to see unless we are extremely rigorous in checking our prejudice.

Anthropologists *do* criticize each other's work, of course—their ferocious personal rivalries are partly responsible for the subjectivity of their judgments—but the question they debate is *whose* set of fossil candidates tells the story of human evolution most accurately, not *whether* fossil proof of the ape-human transition exists. For those who have chosen to devote their lives to exploring exactly how humans evolved from apes, persons who doubt the basic premise are by definition creationists, and hence not to be taken seriously. That there might be no reliable fossil evidence of human evolution is out of the question.

A prestigious outsider, however, has proposed the unthinkable. Solly Zuckerman, one of Britain's most influential scientists and a leading primate expert, is a good scientific materialist who regards the evolution of man from apes as self-evident, but who also regards much of the fossil evidence as poppycock. Zuckerman subjected the *Australopithecines* to years of intricate "biometric" testing, and concluded that "the anatomical basis for the claim that [they] walked and ran upright like man is so much more flimsy than the evidence which points to the conclusion that their gait was some variant of what one sees in subhuman Primates, that it remains unacceptable."

Zuckerman's judgment of the professional standards of physical anthropology was not a generous one: he compared it to parapsychology and remarked that the record of reckless speculation in human origins "is so astonishing that it is legitimate to ask whether much science is yet to be found in this field at all." The anthropologists not surprisingly resented that judgment, which would have left them with no fossils and no professional standing. Wilfred Le Gros Clark performed a rival study that came to more acceptable conclusions, and the consensus of the experts, meaning those who had the most to lose, was that Zuckerman was a curmudgeon with no real feel for the subject. The biometric issues are technical, but the real dispute was a conflict of priorities. Zuckerman's methodological premise was that the first priority of human origins researchers should be to avoid embarrassments like the Piltdown and Nebraska Man fiascos, not to find fossils that they can plausibly proclaim as ancestors. His factual premise was that the variation among ape fossils is sufficiently great that a scientist whose imagination was fired by the desire to find ancestors could easily pick out some features in an ape fossil and decide that they were "pre-human." Granted these two premises, it followed that all candidates for "ancestor" status should be subjected to a rigorous objective analysis, and rejected if the analysis was either negative or inconclusive.

Zuckerman understood that it was probable that none of the ape-like hominid fossils would be able to pass this kind of test, and that as a consequence fossil evidence of human evolution might be limited to specimens like Neanderthal Man that are human or nearly human. The absence of direct evidence for an ape-man transition did not trouble him, because he assumed that the Darwinian model was established for humans as well as other species on logical grounds. Besides, evidence of ancestral relationships is in general absent from the fossil record. That being the case, it should be cause for suspicion rather than congratulation if there were a surfeit of ancestors in the one area in which human observers are most likely to give way to wishful thinking.

Zuckerman's position might have seemed reasonable to persons with no great stake in the question, but one also has to consider the cultural and economic aspects of the situation. The story of human descent from apes is not merely a scientific hypothesis; it is the secular equivalent of the story of Adam and Eve, and a matter of immense cultural importance. Propagating the story requires illustrations, museum exhibits, and television reenactments. It also requires a priesthood, in the form of thousands of researchers, teachers, and artists who provide realistic and imaginative detail and carry the story out to the general public. The needs of the public and the profession ensure that confirming evidence will be found, but only an audit performed by persons not committed in advance to the hypothesis under investigation can tell us whether the evidence has any value as confirmation.

For all these reasons I do not accept the alleged hominid species as independently observed data which can confirm the Darwinian model. I should add, however, that this degree of skepticism is not necessary to make the point that the hominid series cited by Gould is open to question. Some experts in good standing doubt, for example, that *A. Afarensis* and *A. Africanus* were really distinct species, and many deny that there ever was such a species as *Homo habilis*. The most exciting hypothesis in the field right now is the "mitochondrial Eve" theory based upon the molecular clock hypothesis discussed in Chapter Seven, which asserts that modern humans emerged from Africa less than 200,000 years ago. If that hypothesis is accepted, then all the *Homo erectus* fragments found outside of Africa are necessarily outside the ancestral chain, because they are older than 200,000 years.

Still, I am happy to assume *arguendo* that small apes (the *Australopithecines*) once existed which walked upright, or more nearly upright than apes of today, and that there may also have been an intermediate species (*Homo erectus*) that walked upright and had a brain size intermediate between that of modern men and apes. On that assumption there are possible transitional steps between apes and humans, but nothing like the smooth line of development that was proclaimed by Dobzhansky and other neo-Darwinists. We have to imagine what Steven Stanley calls "rapid branching," a euphemism for mysterious leaps, which somehow produced the human mind and spirit from animal materials. Absent confirmation that such a thing is possible, it is reasonable to keep open the possibility that the putative hominid species were something other than human ancestors, even if the fossil descriptions are reliable.

The hominids, like the mammal-like reptiles, provide at most some plausible candidates for identification as ancestors, if we assume in advance that ancestors must have existed. That 130 years of very determined efforts to confirm Darwinism have done no better than to find a few ambiguous supporting examples is significant negative evidence. It is also significant that so much of the claimed support comes from the human evolution story, where subjectivity in evaluation is most to be expected.

The fossils provide much more discouragement than support for Darwinism when they are examined objectively, but objective examination has rarely been the object of Darwinist paleontology. The Darwinist approach has consistently been to find some supporting fossil evidence, claim it as proof for "evolution," and then ignore all the difficulties. The practice is illustrated by the use that has been made of a newly-discovered fossil of a whale-like creature called *Basilosaurus*.

Basilosaurus was a massive serpent-like sea monster that lived during the early age of whales. It was originally thought to be a reptile (the name means "king lizard"), but was soon reclassified as a mammal and a cousin of modern whales. Paleontologists now report that a *Basilosaurus* skeleton recently discovered in Egypt has appendages which appear to be vestigial hind legs and feet. The function these could have served is obscure. They are too small even to have been much assistance in swimming, and could not conceivable have supported the huge body on land. The fossil's discoverers speculate that the appendages may have been used as an aid to copulation.

Accounts of the fossil in the scientific journals and in the newspapers present the find as proof that whales once walked on legs and therefore descended from land mammals. None of these accounts mentions the existence of any unresolved problems in the whale evolution scenario, but the problems are immense. Whales have all sorts of complex equipment to permit deep diving, under-water communication by sound waves, and even to allow the young to suckle without taking in sea water. Step-by-step adaptive development of each one of these features presents the same problems discussed in connection with wings and eyes in Chapter Three. Even the vestigial legs present problems. By what Darwinian process did useful hind limbs wither away to vestigial proportions, and at what stage in the transformation from rodent to sea monster did this occur? Did rodent forelimbs transform themselves by gradual adaptive stages into whale flippers? We hear nothing of the difficulties because to Darwinists unsolvable problems are not important.

Darwin conceded that the fossil evidence was heavily against his theory, and this remains the case today. It is therefore not surprising that Darwinist science has turned its attention to the newly discovered molecular evidence, and claimed that here at last is where conclusive proof of the Darwinian model can be found. We will look at that claim in the next chapter.

Chapter Seven

The Molecular Evidence

BEFORE WE TRY to get any answers out of the molecular evidence, we had better review where we stand. What do we already know, and what do we need to know?

We saw in Chapter Five that it is possible to classify creatures, and that to do so it is necessary to identify the fundamental similarities called homologies that reflect true natural relationship. Both before and after the triumph of Darwinism, classifiers agreed that the relationships so uncovered are not arbitrary but rather express some genuine property of the natural order. Essentialists who rejected evolution thought that the natural groups conformed to the pattern of an archetype, a blueprint existing in some metaphysical realm such as the mind of God. The Darwinists discarded the archetypes and substituted a belief in common ancestors, material beings which existed on earth in the distant past.

The history of life provided by the fossil record is critically important as a test of Darwinism, because the necessary common ances-

86

tors and transitional intermediates are consistently absent from the living world. At the higher levels of the taxonomic hierarchy, today's groups are discontinuous. Every creature belongs to one and only one phylum, class, and order, and there are no intermediates. This is true even of the odd mosaics: the lungfish is a fish, and the duck-billed platypus is a mammal. Pre-Darwinian classifiers cited the absence of intermediates as a conclusive reason for rejecting biological evolution.

Darwinists do not in principle deny the fundamental discontinuity of the living world, but they explain it as being due to the extinction of vast numbers of intermediates that once linked the discrete groups to their remote common ancestors. Some Darwinists like Richard Dawkins have even pointed to present discontinuity with pride, as if it were itself a discovery of Darwinism:

> As long as we stay above the level of the species, and as long as we study only modern animals (or animals in any given time slice . . .) there are no awkward intermediates. If an animal appears to be an awkward intermediate, say it seems to be exactly intermediate between a mammal and a bird, an evolutionist can be confident that it must definitely be one or the other. . . . Indeed, it is important to understand that all mammals—humans, whales, duck-billed platy-puses, and the rest—are exactly equally close to fish, since all mammals are linked to fish via the same common ancestor.

It is, in a way, a blessing, Dawkins added, that the fossil record is imperfect. A perfect fossil record would make classification arbitrary because one category would just blend into another. Many other Darwinists have said the same, and the question for those of us who would like to see proof is whether there is any way to test such statements empirically. In Chapters Four and Six we reviewed the difficulties Darwinists have had in reconciling their premise of past continuity with the inability to identify common ancestors and transitional intermediates in the fossil record, and with the pervasive presence of stasis (the absence of significant change). Today, just as when Darwin first published *The Origin of Species* in 1859, the fossil record as a whole is something that has to be explained away.

Darwinism provided not only a premise of gradual change from ancestors to descendants, but also an explanation of how such

change could create new forms of life and complex biological structures. The mechanism was natural selection of individual
organisms—the most important Darwinian concept—and we reviewed the evidence on this subject in Chapters Two and Three. We
saw there that the hypothesis that natural selection is a major creative force is not well supported empirically, and that Darwinists
have employed the concept as a virtually self-evident logical proposition, something that just must be true. Despite official denials,
Darwinists continue to evoke natural selection this way to account
for whatever innovation or stasis nature happens to have produced.
If new forms appear, the credit goes to creative natural selection; if
old forms fail to change, the conservative force is called stabilizing
selection; and if some species survived mass extinctions while others
perished, it is because the survivors were more resistant to extinction.

Darwinists have consistently said that natural selection was not
the exclusive means of evolution, but they have often been vague
about what else was allowable and how important it could be. They
do not necessarily deny that macromutations have occurred, but
with rare exceptions they vigorously deny that adaptive macromutations could have played an important role in building new forms of
life or complex organs. Saltations or systemic macromutations, by
which all the organs of a body change harmoniously in a single
generational leap, are out of the question as virtual genetic miracles.
Some neutral evolution, or "genetic drift," is clearly possible. Darwinists believe that variations arise by chance, and they can spread
by chance, but the most logically rigorous Darwinists have insisted
that variants must soon pass the test of natural selection or vanish.

This position is a natural inference from the basic principles of
Darwinism. Even very small changes must have a significant impact
upon reproductive success if natural selection is to perform the
necessary wonders of craftsmanship. Recall how Dawkins explained
the evolution of the wing, for example. He argued that the first
(probably imperceptible) micromutation in that direction must have
conferred some small selective advantage, perhaps by preventing
the creature from breaking its neck in a fall. If creatures can vary
substantially without any significant effect upon survival or reproductive success, however, natural selection cannot get to work until

the creature is pretty far along in growing wings. "Pan-selectionism"—the doctrine that natural selection preserves or eliminates even minute variations—is a logical consequence of the assumption that natural selection can build complex biological structures with only micromutations for raw material.

Natural selection operates directly upon the characters of the phenotype[1] that function in the environment, but by logical extension it must have a similar effect upon the genetic material that contains the information that produces those characters in the reproductive process. The authoritative Ernst Mayr therefore announced in 1963, as the molecular revolution was beginning, that "I consider it exceedingly unlikely that any gene will remain selectively neutral for any length of time."

The purpose of this review has been to clarify what we would have to find in the molecular evidence, or any other body of new evidence, before we would be justified in concluding that Darwinism is probably true. We would need to find evidence that the common ancestors and transitional intermediates really existed in the living world of the past, and that natural selection in combination with random genetic changes really has the kind of creative power claimed for it. It will not be enough to find that organisms share a common biochemical basis, or that their molecules as well as their visible features can be classified in a pattern of groups within groups. The important claim of Darwinism is not that relationships exist, but that those relationships were produced by a naturalistic process in which parent species were gradually transformed into quite different descendant forms through long branches (or even thick bushes) of transitional intermediates, without intervention by any Creator or other non-naturalistic mechanism. If Darwinism so defined is false then we do not have any important scientific information about how life arrived at its present complexity and diversity, and we cannot turn ignorance into information by calling it evolution.

With the agenda of questions clarified, we go now to the evidence

---

[1] The "phenotype" refers to the visible features of an organism, or more precisely to the detectable expression of the interaction between the genotype and the environment. The genotype is the invisible package of genes that directs the growth of the phenotype in the reproductive process.

to see what it tells us about the power of natural selection and about whether the existence of common ancestors and intermediates can be empirically confirmed.

BECAUSE OF enormous advances in biochemistry, it has become possible to compare not just the visible features of organisms, but also their molecules. The principle components of the biological cell include the proteins, which govern the essential biochemical processes, and the nucleic acids (the famous DNA and RNA), which direct the synthesis of proteins. The structure and composition of these immensely complex molecules is now partly understood, and so the proteins and nucleic acids of various kinds of creatures can be compared and their differences precisely quantified.

Each protein molecule, for example, consists of a long chain of amino acids in a specific sequence, analogous to the way a sentence is composed of a sequence of letters and spaces in a particular order. Amino acids are simpler organic compounds, 20 of which can be combined in various ways to make proteins. A particular kind of protein (like hemoglobin) that is found in a great variety of species will differ slightly or not so slightly in its amino acid sequences from species to species. The difference can be quantified by aligning the sequences and counting the number of positions at which the amino acids differ. If there are a total of 100 positions, and the amino acids are the same at 80 of them and different at 20, then the biochemist can say that the degree of divergence is 20 per cent.

Comparable techniques can be employed to measure the divergence in the molecular sequences of DNA and RNA molecules. As a result, biochemists have found that it is possible to classify species and larger groups by their degree of similarity at the molecular level. The validity of the classifications so obtained is a controversial subject. Not all molecules suggest the same pattern of relationships, and in some cases molecular classifications differ from traditional classifications. Moreover, there seems to be no necessary relationship between the degree of molecular difference between two species and any differences in tangible characteristics. All frog species look pretty much alike, for example, but their molecules differ as much as those of mammals, a group which contains such fantastically diverse forms as the whale, the bat, and the kangaroo.

Despite these difficulties, many scientists consider molecular classification to be not only possible, but, in principle, more objective than classification based on visible characteristics. Molecular studies have also produced claims having important philosophical implications, particularly on the sensitive topic of human evolution, because by some molecular measurements chimps are much more similar to humans than they are to other non-human primates. This degree of similarity may call the importance of molecular comparisons into question, because it does little to explain the profound *dis*similarities between humans and animals of any kind. Evidently the information content of the human genetic system is significantly different from that of apes, even though the arrangement of chemical "letters" looks almost the same. This point is lost on some Darwinists. In *Blueprints: Solving the Mystery of Evolution*, Maitland Edey and Donald Johanson casually declare that: "Although humans may look entirely different from chimpanzees and gorillas, those differences are superficial. Where it counts—in their genes—all three are ninety-nine percent identical." There is a lot of philosophy packed in that phrase "where it counts."

Because Darwinists take for granted that "relationship" is equivalent to common ancestry, they assume that the molecular classifications confirm the "fact of evolution" by confirming the existence of something which by definition is caused by evolution. They also tend to assume that the particular relationships determined by taxonomists were "predicted" by Darwin's theory. When these fallacious assumptions are made, it seems that a "99 per cent" molecular similarity between men and apes confirms Darwinism decisively.

The misunderstanding is fundamental. Darwin did not invent classification or reform its practice. His contribution was to provide an explanation in materialistic terms of how the categories came about and why the classifiers were right in their instinct that the "types" are real natural entities and not arbitrary sorting systems (such as a library uses for books). Pre-Darwinian classifiers also were aware that humans are physically very much like the anthropoid apes. That is why the creationist Linnaeus, the father of taxonomy, unhesitatingly included humans among the primates. The genetic similarity confirms Linnaeus, not Darwin. It tells us once again that apes and humans are remarkably similar in some ways, just as they are remarkably different in others, but it does not tell us how either the similarities or the dissimilarities came to exist.

One thing the molecular evidence does confirm is that the groups of the natural order are isolated from each other, which is to say they are not connected by any surviving intermediate forms. A protein called cytochrome c which is found in a great variety of species has been studied extensively. A standard reference table shows the percent sequence divergence between the cytochrome c of a particular bacterium and a wide variety of more complex organisms, including mammals, birds, reptiles, amphibians, fish, insects, and angiosperms (plants). The sequence divergence from the bacterial form ranges from 64 percent (rabbit, turtle, penguin, carp, screw worm) to 69 (sunflower). If the comparison is restricted to animals, from insects to man, the range is only from 64 to 66.

Judged by cytochrome c comparisons, sesame plants and silkworms are just about as different from bacteria as humans are. In fact, every plant and animal species is approximately the same molecular distance from any bacterial species, and there is no surviving trace of any intermediates that might have filled the "space" between single-celled and multicellular life. If the molecules evolved gradually to their present form, then intermediates must over time have filled that space, but comparing present-day molecules cannot tell us whether these transitional forms ever existed.

Another result of molecular studies has been to reveal that there are a greater number of fundamental divisions in the living world than had previously been recognized. A biochemist named Woese compared the "RNA sequences" in a wide variety of organisms. RNA is a very important macromolecule which in all kinds of living organisms helps to form proteins. Before Woese published his results everyone had assumed that the fundamental division in nature was between prokaryotes (bacteria) and eukaryotes (all plants and animals). The difference between the two is one of fundamental cell structure. The prokaryote cell has no true nucleus, and the eukaryote cell has a nucleus enclosed by its own membrane. Woese and his colleagues showed that the prokaryote kingdom includes two entirely distinct kinds of bacteria, as different from each other at the molecular level as either is from the eukaryotes.

This means that there are three primary divisions of the living world (in terms of cellular construction) rather than two. Woese renamed the more conventional prokaryotes the eubacteria, and called the new kingdom the archaebacteria. The archaebacteria all favor what we would consider unusual lifestyles: one anaerobic group can manufacture methane gas, another likes salt-saturated environments that kill nearly everything else, and a third prefers extra high temperature settings like hot sulphur springs. The prefix "archae" means "old." Woese chose it because he speculated that a group favoring such extreme environments might have been suited to conditions thought to prevail on the early earth. That might suggest that archaebacteria are ancestral to eubacteria, but the two bacterial kingdoms are so fundamentally different from each other that neither could have evolved from the other. They are separated by an immense molecular distance, (and plenty of more tangible characteristics) with nothing in between.

Biochemists assume that the three cellular kingdoms evolved from a single common ancestor, because the alternative of supposing an independent origin of life two or more times presents still greater difficulties. This common ancestor is merely hypothetical, as are the numerous transitional intermediate forms that would have to connect such enormously different groups to the ancestor. From a Darwinist viewpoint all these hypothetical creatures are a logical necessity, but there is no empirical confirmation that they existed.

That brings us to the second major question discussed in the introductory paragraphs to this chapter. Darwinian theory insists that natural selection is a creative force of immense power, which preserves the slightest favorable variations and spreads them throughout a breeding population so that further favorable micromutations can accumulate and produce new characteristics of formidable complexity, such as wings and eyes. We have already seen that the hypothesis of creative natural selection lacks experimental support, and that it is disconfirmed by the fossil record. The molecular evidence adds further doubt, because of the previously described phenomenon of molecular *equidistance*.

Consider a small part of what supposedly happened in the mammal line, for example, after this group "split" from its hypothetical last common ancestor with modern reptiles. A number of other splits followed, and one of these new lines went towards the water and, after an almost inconceivable set of adaptive changes became the first whale. A different line took to the trees and caves, learned stepby-step to fly, and developed a "sonar" navigation system as a substitute for sight. The experiences of the two lines were as different as their eventual forms, but it now appears that all these differences had no effect on the rate of change in cytochrome c and various other molecules. When homologous molecules of contemporary whales and bats are compared, they are each at roughly equal molecular distances from comparison molecules of any modern reptile like the snake, which by hypothesis had been taking its own separate path to its present form. For reasons that will shortly be explained, this astonishing phenomenon came to be know as the "molecular clock."

How could such a coincidence happen? It could happen if the rate of molecular change was independent of what was going on in the phenotypes, and unaffected by natural selection. In other words, if molecular evolution occurred at clock-like rates it must have been the product of regularly-occurring mutations that were not greatly affected by the environmental conditions that are presumed to have produced rapid change and lengthy stasis in the phenotypes. This is the essential premise of the neutral theory of molecular evolution, whose leading advocate is Motoo Kimura.

Many Darwinists at first found the neutral theory incredible. Mutations occur in individual organisms, and according to Darwinist theory they spread throughout a population through natural selection. How could a neutral mutation (which by definition confers no reproductive advantage) spread to become a characteristic of the entire species? And how could an organism undergo significant functional changes in its biochemistry without any effect on its fitness?

The neutralists had answers to all the objections. There are many variations in molecular sequences that do not appear to have any functional impact upon the organism. For example, there are redundant DNA sequences that do not code for proteins, and the DNA language contains synonyms, meaning variant sequences that convey the same "message." To the extent that molecular mutations do not have any significant functional effect no one should expect natural selection to guide molecular evolution.

Neutral mutations spread randomly as they happen to occur and as they happen to be passed on to descendants. A particular mutation can become fixed (characteristic of the entire breeding population) simply as a result of surviving a long continuous process of random sampling in which alternative forms were eliminated. Absent special circumstances the neutral theory predicts a high degree of heterozygosity—the co-existence of variant genetic forms called alleles—in contemporary populations. Natural selection would tend to eliminate the less advantageous forms. Neutral evolution, by definition, does not discriminate, and in the real world, greater heterozygosity than selection would seemingly allow is often found.

So far the explanation is logically sound, although Kimura conceded that it depends upon assumptions about past mutation rates, population sizes, and selective effects that cannot be tested independently. Kimura put himself on slippery ground, however, when he argued that the selective effect of a *functional* genetic change depends entirely upon whether it actually affected survival and reproduction. In his own words:

> The neutral theory . . . does not assume that neutral genes are functionless but only that various alleles may be equally effective in promoting the survival and reproduction of the individual. . . .Some criticisms of the neutral theory arise from an incorrect definition of "natural selection." The phrase should be applied strictly in the Darwinian sense: natural selection acts through—and must be assessed by—the differential survival and reproduction of the individual. The mere existence of detectable functional differences between two molecular forms is not evidence for the operation of natural selection, which can be assessed only through investigation of survival rates and fecundity.

Kimura's argument is merely another attempt to rescue the natural selection hypothesis from potential falsification by redefining it as a tautology. If fitness is determined only by the brute fact of survival and reproductive success, then there is no effective difference between neutral and selective evolution. Both illustrate the survival of the fittest, the fittest being those who survive.

Neutralists can also explain how a large amount of neutral molecular evolution can coexist with selective evolution of phenotypes. There are so many molecular mutations that, conceivably, a small percentage might produce enough favorable mutations for natural selection to use in building complex adaptive structures. On that (unverifiable) assumption, selectionist evolution of phenotypes is still possible even if most molecular changes are selectively neutral. Kimura added that natural selection is important in the neutral theory in its negative, conservative sense. There is evidence that variation occurs most frequently at molecular sites which do not control functions critical to the life process, and less frequently at "constrained" sites, where alterations could adversely affect important functions. At the molecular level, the effect of natural selection is therefore mainly to prevent change.

Whatever its effect on other issues, the molecular evidence does nothing to provide the hypothesis of creative natural selection with the empirical confirmation it so badly needs. Natural selection is a force for building adaptive complexity only when it is formulated as a tautology or as a logical deduction unconnected to any empirically verifiable reality. Whenever natural selection is actually observed in operation, it permits variation only within boundaries and operates as effectively to preserve the constraining boundaries as it does to permit the limited variation. The hypothesis that natural selection has the degree of creative power required by Darwinist theory remains unsupported by empirical evidence.

The neutralist-selectionist argument never needs to be settled, because selectionist explanations may have an advantage with respect to one set of data and neutralist explanations with another. Both sides are Darwinists in the only important sense: they assume that natural selection shaped the phenotypes, and that random genetic change provided the raw material of evolution. The neutral theory was proposed not to challenge Darwinism, but rather as an imaginative way to reconcile some very surprising data with the essential elements of Darwin's theory. Far from posing a danger, it greatly increased Darwinism's explanatory power.

The concept of neutral evolution at clock-like rates implied that molecular biologists had discovered a powerful tool for dating macroevolutionary events. If we assume common ancestors for today's living groups—connected to the present world by long lines of vanished intermediates—then it is possible to estimate the amount of time that has passed since any two species "split" from their last common ancestor. Because changes seem to accumulate in homologous molecules in diverse species at roughly constant rates, all that is necessary is to "calibrate the molecular clock" in one species against the date of some evolutionary transition estimated from the fossil record. Equivalent molecules in other species should theoretically have been changing at the same rate, and so by comparing the appropriate molecules of any two species the biochemist can determine how long ago they split from their assumed common ancestor.

The molecular clock was put to effective use by Berkeley's Allan Wilson and Vincent Sarich, and had an important impact upon accepted notions of human descent. Anthropologists relying upon fossil evidence had estimated that the ape and human lineages had split at least 15 million years ago, but the molecular calculations supported a period of between 5 and 10 million years. A date of around 7 million years has come to be widely accepted, in large part because of the influence of the molecular data. More recently, Wilson and others have studied descent within the human species by analyzing mitochondrial DNA, which is passed only in the female line, from mother to daughter. Their conclusion is that all contemporary humans are descendants of a woman who lived in Africa less than 200,000 years ago. Some anthropologists do not accept this conclusion, however, in part because it implies that all the *Homo erectus* fossils found outside of Africa that are older than 200,000 years could not be in the line of descent leading to modern humans. Conflict is developing between fossil experts and molecular biologists over which discipline has the authority to settle disputes over the course of human evolution.

Darwinists regularly cite the molecular clock findings as the decisive proof that "evolution is a fact." The clock is just the kind of thing that intimidates non-scientists: it is forbiddingly technical, it seems to work like magic, and it gives impressively precise numerical figures. It comes from a new branch of science unknown to Darwin, or even to the founders of the neo-Darwinian synthesis, and the scientists say that it confirms independently what they have been telling us all along. The show of high-tech precision distracts attention from the fact that the molecular clock hypothesis *assumes* the validity of the common ancestry thesis which it is supposed to confirm.

What the molecular evidence actually provides is a restatement of the argument from classification. The molecular relationships that have been reported so far are generally (but not entirely) consistent with classifications based on visible features. Divergence dates calculated from the molecular relationships are also said to be roughly consistent with estimates of the first appearance of new groups according to the fossil evidence.[2] Like the relationships determined from visible characteristics, the molecular relationships could have come about by divergence from common ancestors, if the ancestors ever existed.

To a Darwinist, that possibility is more than just evidence for evolution. It *is* evolution, because to Darwinists relationship means evolutionary relationship. And the fact carries with it all the necessary corollaries, including whatever creative power has to be attributed to natural selection to make it possible for simple ancestors to change into complex descendants. As a consequence of this logic, Darwinists consider it perverse that anyone familiar with the molecular evidence would doubt "evolution"—meaning the gradual, naturalistic development of all life forms by descent with modification all the way from prokaryotes to humans.

If variations in molecules were the only thing that needed to be explained, there would be no reason to doubt that neutral mutations can accumulate and cause a pattern of molecular relationships. The trouble is that the molecules had to be embodied in organisms, which had to be evolving from ancestral to descendant forms along with the molecules. The common ancestors and transitional links are still only theoretical entities, conspicuously absent from the fossil record even after long and determined searching.

More important still, science knows of no natural mechanism capable of accomplishing the enormous changes in form and function required to complete the Darwinist scenario. A theory that

---

[2] In this chapter I am taking the neutral theory and the molecular clock data at face value, but I should note that the whole subject is currently embroiled in complex controversy. According to a recent review article by Roger Lewin, "The theory that we can date the birth of new species by charting the steady accumulation of mutations over evolutionary time is in serious trouble." It seems that the data are too even for a selectionist interpretation, and not even enough for a neutralist explanation. According to Allan Wilson, "many biologists who make mathematical models of the evolutionary process are coming to believe many of the mutations accumulated during molecular evolution are not neutral. They argue that instead of proceeding smoothly, molecular evolution might be characterized by long periods of inactivity punctuated by bursts of change. If they are right, the challenge of finding an explanation for the molecular clock phenomenon grows." About all that can be said for now is that a pattern of relationships exists at the molecular level which is roughly consistent with the relationships determined by visible features, and which could have come about by some combination of variable and constant-rate evolution.

explains only changes that have no important functional effects does nothing to solve the real mystery of evolution, which is how the marvelous molecular structures could have evolved in the first place, and how a (relatively) simple cell could change into a complex plant or animal. On the contrary, molecular biology adds to the difficulty by revealing that the molecules themselves are pieces of intricate machinery that require the cooperation of numerous complex parts to carry out their functions. The hemoglobin molecule, for example, is so complex in its architecture and function that it is sometimes called the "molecular lung." The difficulties of explaining how living structures could evolve by mutation and selection grow greater as each additional level of complexity is uncovered.

The molecular evidence therefore fails to confirm either the reality of the common ancestors or the adequacy of the Darwinist mechanism. In fact, testing Darwinism by the molecular evidence has never even been attempted. As in other areas, the objective has been to find confirmation for a theory which was conclusively presumed to be true at the start of the investigation. The true scientific question—Does the molecular evidence as a whole tend to confirm Darwinism when evaluated without Darwinist bias?—has never been asked.

In this chapter we have reviewed evidence concerning similarities and differences in the proteins and nucleic acids that are among the most fundamental components of all living organisms. The question remains how these complex molecular structures came into existence in the first place. That brings us to our next subject, which is the origin of life itself.

Chapter Eight

Prebiological Evolution

WHEN THE SUPREME COURT struck down the Louisiana law requiring balanced treatment for creation-science, Justice Antonin Scalia dissented from the decision because he thought that "The people of Louisiana, including those who are Christian fundamentalists, are quite entitled . . . to have whatever scientific evidence there may be against evolution presented in their schools." Stephen Jay Gould was baffled that a jurist of Scalia's erudition (he had held professorships at several major universities) would entertain the absurd notion that fundamentalists could have scientific evidence against evolution. Gould went looking in Scalia's opinion for an explanation, and found it in various sentences implying that evolution is a theory about the origin of life.

In an article correcting "Justice Scalia's Misunderstanding," Gould tried to set the matter straight. Evolution, he wrote, "is not the study of life's ultimate origin, as a path toward discerning its deepest meaning." Even the purely scientific aspects of life's first appearance on earth belong to other divisions of science, because "evolution" is merely the study of how life changes once it is already in existence. Because he misunderstood the strictly limited subject matter of evolution, Scalia had tumbled into the misunderstanding that it is possible to have rational objections to the doctrines of evolutionary science.

In fact, Justice Scalia used the general term "evolution" exactly as scientists use it—to include not only *biological* evolution but also *prebiological* or chemical evolution, which seeks to explain how life first evolved from nonliving chemicals. Biological evolution is just one major part of a grand naturalistic project, which seeks to explain the origin of everything from the Big Bang to the present without allowing any role to a Creator. If Darwinists are to keep the Creator out of the picture, they have to provide a naturalistic explanation for the origin of life.

Speculation about prebiological evolution began to appear as soon as *The Origin of Species* had made its impact, with Darwin's "German Bulldog" Ernst Haeckel taking the leading role at first. Darwin himself made a famous contribution to the field in an 1871 letter:

> It is often said that all the conditions for the first production of a living organism are now present, which could ever have been present. But if (and oh! what a big if!) we could conceive in some warm little pond, with all sorts of ammonia and phosphoric salts, lights, heat, electricity, etc. present, that a protein compound was chemically formed ready to undergo still more complex changes, at the present day such matter would be instantly devoured or absorbed, which would not have been the case before living creatures were formed.

Robert Shapiro observed in 1986 that Darwin's offhand speculation "is remarkably current today, which is a tribute either to his foresight or our lack of progress." A generation ago the field of prebiological evolution seemed on the brink of spectacular success; today it is just about where Darwin left it.

The basic difficulty in explaining how life could have begun is that all living organisms are extremely complex, and Darwinian selection cannot perform the designing even in theory until living organisms already exist and are capable of reproducing their kind.

A Darwinist can imagine that a mutant rodent might appear with a web between its toes, and thereby gain some advantage in the struggle for survival, with the result that the new characteristic could spread through the population to await the arrival of further mutations leading eventually to winged flight. The trouble is that the scenario depends upon the rodent having offspring that inherit the mutant characteristic, and chemicals do not produce offspring. The challenge of chemical evolution is to find a way to get some chemical combination to the point where reproduction and selection could get started.

The field achieved its greatest success in the early 1950s when Stanley Miller, then a graduate student in the laboratory of Harold Urey at the University of Chicago, obtained small amounts of two amino acids by sending a spark through a mixture of gases thought to simulate the atmosphere of the early earth. Because amino acids are used in building proteins, they are sometimes called the "building blocks of life." Subsequent experiments based on the Miller-Urey model produced a variety of amino acids and other complex compounds employed in the genetic process, with the result that the more optimistic researchers concluded that the chemicals needed to construct life could have been present in sufficient abundance on the early earth.

The Miller-Urey experiment partially validated a theoretical model proposed by Alexander Oparin and J. B. S. Haldane in the 1920s. The Oparin-Haldane model postulated first that the early earth had a "reducing" atmosphere made up of gases like methane, hydrogen, and ammonia, with little or no free oxygen. Second, into this atmosphere came various forms of energy, like the electric sparks in the Miller-Urey apparatus, forming the essential organic compounds. Third, in Haldane's words, these compounds "must have accumulated until the primitive oceans reached the consistency of hot dilute soup." Haldane's metaphor caught the journalistic imagination and the "prebiotic soup" has become an element of scientific folklore, presented to the public in books and museum exhibits as the known source of early life. The fourth element in the theory was the most important and also the most mysterious: somehow life emerged from the prebiotic soup.

The limited success of the Miller-Urey experiment occurred in the years leading up to the Darwinian Centennial celebrations in

1959. This was the height of neo-Darwinist triumphalism, just when the literally smashing debut of atomic energy had made it seem that all mysteries would yield to the power of scientific investigation. In that climate of opinion, the experiment appeared to have created life by a technique reassuringly similar to that employed by Dr. Frankenstein in the movies. The 1980s have been a period of skeptical reassessment, however, during which specialists called into question each of the four elements in the Oparin-Haldane scenario.

Geochemists now report that the atmosphere of the early earth probably was not of the strongly reducing nature required for the Miller-Urey apparatus to give the desired results. Even under ideal and probably unrealistic conditions, the experiments failed to produce some of the necessary chemical components of life. Perhaps the most discouraging criticism has come from chemists, who have spoiled the prebiotic soup by showing that organic compounds produced on the early earth would be subject to chemical reactions making them unsuitable for constructing life. In all probability, the prebiotic soup could never have existed, and without it there is no reason to believe that the production of small amounts of some amino acids by electrical charge in a reducing atmosphere had anything to do with the origin of life.

Although these objections to the significance of the Miller-Urey results are important, for present purposes I prefer to disregard them as a distraction from the main point. Let us grant that, one way or another, all the required chemical components were present on the early earth. That still leaves us at a dead end, because there is no reason to believe that life has a tendency to emerge when the right chemicals are sloshing about in a soup. Although some components of living systems can be duplicated with very advanced techniques, scientists employing the full power of their intelligence cannot manufacture living organisms from amino acids, sugars, and the like. How then was the trick done before scientific intelligence was in existence?

The simplest organism capable of independent life, the prokaryote bacterial cell, is a masterpiece of miniaturized complexity which makes a spaceship seem rather low-tech. Even if one assumes that something much simpler than a bacterial cell might suffice to start Darwinist evolution on its way—a DNA or RNA macromolecule, for example—the possibility that such a complex entity could assemble itself by chance is still fantastically unlikely, even if billions of years had been available.

I won't quote figures because exponential numbers are unreal to people who are not used to them, but a metaphor by Fred Hoyle has become famous because it vividly conveys the magnitude of the problem: that a living organism emerged by chance from a pre-biotic soup is about as likely as that "a tornado sweeping through a junkyard might assemble a Boeing 747 from the materials therein." Chance assembly is just a naturalistic way of saying "miracle."

A scientific explanation of this miracle is not absolutely necessary, because *in extremis* Darwinists can handle the problem with philo-sophical argument. Life obviously exists, and if a naturalistic pro-cess is the only conceivable explanation for its existence, then the difficulties must not be as insuperable as they appear. Even the most discouraging aspects of the situation can be turned to advantage when they are viewed with the eye of faith. For example, life seems to have existed in cellular form nearly four billion years ago, per-haps as soon as the earth had sufficiently cooled. That means that the emergence of the first self-replicating molecules, and the subse-quent evolution of all the machinery of the cell, had to occur within a brief period of geological time. Far from being discouraged by the limited time available, Carl Sagan drew the conclusion that life was likely to have evolved on other planets as well. His reasoning was that the spontaneous origin of life must be relatively easy, since it happened so quickly on the early earth.

For those not so easily satisfied, the cosmological "anthropic prin-ciple" is available to tame the unfavorable odds. This principle starts with the existence of observers—ourselves -and works backwards. If the circumstances required for life to evolve had not existed we would not be here to comment upon the matter. Those circum-stances may seem very unlikely given our limited knowledge, but we have no way of knowing how many universes there are, or may have been. In an infinity of time and space even the most unlikely event must happen at least once, and we necessarily exist in the corner of reality where the particular set of coincidences necessary for our existence happened to occur.

Richard Dawkins, who has Darwin's own facility for turning a liability into an asset, has even argued that the improbability of the origin of life scenarios is a point in their favor. He reasons that "An apparently (to ordinary human consciousness) miraculous theory is *exactly* the kind of theory we should be looking for in this particular matter of the origin of life." This is because "evolution has equipped our brains with a subjective consciousness of risk and improbability suitable for creatures with a lifetime of less than one century."

Dawkins is actually *encouraged* by the failure of scientists to duplicate the spontaneous generation of life in their laboratories. After all, scientists can't duplicate biological macroevolution either. If making life were easy enough that scientists could do it, then nature would have caused life to originate spontaneously on earth many times, as well as on planets within radio range. As it appears that this did not happen, failure to duplicate the origin of life in the laboratory is just what Darwinist theory would lead us to expect.[1]

When it becomes necessary to rely on arguments like that one, the experimental work must be going very badly. For those who prefer to address the problem with scientific methodology instead of rhetorical virtuosity, a way must be found to extend the concept of evolution to a level prior to the molecules of the genetic system. In contemporary organisms, DNA, RNA, and proteins are mutually interdependent, with DNA storing the genetic information and copying it to RNA, RNA directing the synthesis of proteins, and proteins carrying on the essential chemical work of the cell. An evolutionary scenario must assume that this complex system evolved from a much simpler predecessor, probably employing at first only one of the three major constituents. Which came first, the nucleic acids (DNA or RNA) or the proteins? And how did the first living molecule function and evolve in the absence of the others?

Those questions define the agenda for the field of chemical evolution, where several scenarios are competing for attention. I will describe the leading candidates only briefly, because the subject is well covered in other books and there is widespread agreement that no theory has obtained any substantial experimental confirmation.

For some time the most popular contender has been the "naked gene" or "RNA first" hypothesis, based on the premise that life began when an RNA molecule somehow managed to synthesize

---

[1] If readers suspect that Dawkins was not being serious when he advanced this argument, they are probably correct. He concluded the passage with the following sentence: "Having said all this I must confess that, because there is so much uncertainty in the calculations, if a chemist *did* succeed in creating spontaneous life I would not be disconcerted!"

itself from the organic compounds of the prebiotic soup. RNA is the most likely candidate for the first component of the genetic system because it not only acts as the carrier of genetic information in its "messenger" role, but it also is capable of catalyzing some chemical reactions in the manner of proteins. With this versatility it is conceivable that RNA might have carried on the essential functions of life in a primitive manner until true DNA and proteins could evolve.

Conceivable is a long way from probable or experimentally verifiable, of course. In previous chapters we saw that there is no evidence that Darwinian selection is a sufficiently powerful designing force to transform a molecule or a cell into an abundance of complex plants and animals, even given a few billion years. Origin of life chemists take universal biological Darwinism for granted, but they can identify plenty of problems with the proposition that a self-replicating RNA molecule could have evolved from organic compounds on the early earth. The obstacles to prebiotic RNA synthesis were reviewed in 1989 in a lengthy article by G. F. Joyce in *Nature*. Joyce concluded that RNA is "not a plausible prebiotic molecule, because it is unlikely to have been produced in significant quantities on the primitive earth." As with other once-promising models of prebiological evolution, the "RNA-first" theory cannot survive detailed examination.

Joyce surmised that RNA itself would have had to have evolved from some simpler genetic system which is no longer in existence. An imaginative idea about what a prebiotic genetic system might have been like has been proposed by A. G. Cairns-Smith, most recently in a charming book titled *Seven Clues to the Origin of Life*. Bizarre as the idea may appear at first, or even upon reflection, Cairns-Smith thinks that clay crystals have qualities that might make possible their combination into a form of pre-organic mineral life. According to Darwinist assumptions, natural selection would then favor the more efficient clay replicators, preparing the way for an eventual "genetic takeover" by organic molecules that had evolved because of their increasing usefulness in the pre-organic process.

The imagination involved in the mineral origin of life thesis is impressive, but for my purpose it is sufficient to say that it is altogether lacking in experimental confirmation. According to the biochemist Klaus Dose, "This thesis is beyond the comprehension of all biochemists or molecular biologists who are daily confronted with the experimental facts of life." That would ordinarily be more than enough reason to discard a theory, but many scientists still take the idea of a mineral origin of life seriously because there is no clearly superior competitor.

There are other possibilities, including a "protein first" scenario that had appeared to be going out of fashion, but which may make a comeback due to the devastating criticism the RNA rival has recently suffered. In fact, the absence of experimental support for any one theory leaves the door open for just about any speculation other than creationism. A general review of prebiological evolutionary theories in 1988 by Klaus Dose concluded that "At present all discussions on principal theories and experiments in the field either end in stalemate or in a confession of ignorance." Gerald Joyce's 1989 review article ended with the somber observation that origin of life researchers have grown accustomed to a "lack of relevant experimental data" and a high level of frustration.

Prospects for experimental success are so discouraging that the more enterprising researchers have turned to computer simulations that bypass the experimental roadblocks by employing convenient assumptions. An article in *Science* in 1990 summarized the state of computer research into "spontaneous self-organization," a concept based upon the premise that complex dynamical systems tend to fall into a highly ordered state even in the absence of selection pressures. This premise may seem to contradict the famous Second Law of Thermodynamics, which says that ordered energy inevitably collapses into disorder or maximum "entropy." There is reason to believe, however, that in a local system (the earth) which takes in energy from outside (the sun), the second law permits some kinds of spontaneous self-organization to occur. For example, ordered structures like snowflakes and crystals are common. More to the point, most scientists assume that *life* originated spontaneously and thereafter evolved to its present state of complexity. This could not have happened unless powerful self-organizing tendencies were present in nature.

Starting from assumptions like that, scientists can design computer models that mimic the origin of life and its subsequent evolution. Whether the models have any connection to reality is another question. According to *Science*, "Advocates of spontaneous organization are quick to admit that they aren't basing their advocacy on empirical data and laboratory experiments, but on abstract mathematics and novel computer models." The biochemist G. F. Joyce commented: "They have a long way to go to persuade mainstream biologists of the relevance [of this work]."

Assuming away the difficult points is one way to solve an intractable problem; another is to send the problem off into space. That was the strategy of one of the world's most famous scientists, Francis Crick, co-discoverer of the structure of DNA. Crick is thoroughly aware of the awesome complexity of cellular life and the extreme difficulty of explaining how such life could have evolved in the time available on earth. So he speculated that conditions might have been more favorable on some distant planet.

That move leaves the problem of getting life from the planet of origin to earth. First in a paper with Leslie Orgel, and then in a book of his own, Crick advanced a theory he called "directed panspermia." The basic idea is that an advanced extraterrestrial civilization, possibly facing extinction, sent primitive life forms to earth in a spaceship. The spaceship builders couldn't come themselves because of the enormous time required for interstellar travel; so they sent bacteria capable of surviving the voyage and the severe conditions that would have greeted them upon arrival on the early earth.

What kind of scientific evidence supports directed pan-spermia? Crick wrote that if the theory is true, we should expect that cellular microorganisms would appear suddenly, without evidence that any simpler forms preceded them. We should also expect to find that the early forms were distantly related but highly distinct, with no evidence of ancestors because these existed only on the original planet. This expectation fits the facts perfectly, because the archaebacteria and eubacteria are at the same time too different to have evolved from a common ancestor in the time available, and yet also too similar (sharing the same genetic language) not to have a common source somewhere. Those who are tempted to ridicule directed pan-spermia should restrain themselves, because Crick's extraterrestrials are no more invisible than the universe of ancestors that earth-bound Darwinists have to invoke.

Crick would be scornful of any scientist who gave up on scientific research and ascribed the origin of life to a supernatural Creator. But directed pan-spermia amounts to the same thing. The same limitations that made it impossible for the extraterrestrials to journey to earth will make it impossible for scientists ever to inspect their planet. Scientific investigation of the origin of life is as effectively closed off as if God had reserved the subject for Himself.

When a scientist of Crick's caliber feels he has to invoke undetectable spacemen, it is time to consider whether the field of prebiological evolution has come to a dead end. And yet, despite the absence of experimental success, many scientists remain confident that the problem will be solved in the foreseeable future. To understand that confidence, we need to examine the most important intellectual question in the field—the way scientists define the "life" whose origin they are trying to discover.

In *Seven Clues to the Origin of Life*, A. G. Cairns-Smith explains the Darwinist conception of life which underlies the field of prebiological evolution. "Life is a product of evolution," he writes, and the indispensable element in evolution is natural selection. This means that the purpose of a living thing "is to survive, to compete, to reproduce its kind against the odds." The goal of prebiological science therefore, is to find (or at least to imagine) the simplest combination of chemicals that might be capable of competing and reproducing, so that natural selection can begin its work. In this view, natural selection is not just something that happens to life; it is the defining characteristic of life.

When "life" is *defined* as matter evolving by natural selection, there is every reason to be confident of finding an evolutionary explanation for its origin. If Darwin really explained in 1859 how all the complex and diverse forms of life can evolve from a single microorganism, then surely our much more advanced science will not long be stymied at the final step. But what if Darwin was wrong, and natural selection doesn't have the fantastic creative power Darwinists credit it with? In that case prebiological science has misconceived the problem, and its efforts are as doomed to futility as the efforts of medieval alchemists to transform lead into gold.

The Darwinistic definition of life is Cairns-Smith's philosophical preference. When he describes what he actually sees, however, he tells of something very different:

After all what impresses us about a living thing is its in-built ingenuity, its appearance of having been designed, thought out—of having been put together with a purpose. . . . The singular feature is the [enormous] gap between the simplest conceivable version of organisms as we know them, and components that the Earth might reasonably have been able to generate. . . . But the real trouble arises because too much of the complexity seems to be necessary to the whole way in which organisms work.

Cairns-Smith also describes the "messages" contained in the genetic information stored in the "library" of each cell's DNA, which are transcribed and translated to direct the synthesis of proteins. His language is entirely typical of others who write about this subject: practically all stress the appearance of design and purpose, the immense complexity of the simplest cell, and the apparent need for many complex components to work together to sustain life. Everyone uses the vocabulary of intelligent communication to describe protein synthesis: messages, programmed instructions, languages, information, coding and decoding, libraries.

Why not consider the possibility that life is what it so evidently seems to be, the product of creative intelligence?[2] Science would not come to an end, because the task would remain of deciphering the languages in which genetic information is communicated, and in general finding out how the whole system *works*. What scientists would lose is not an inspiring research program, but the illusion of total mastery of nature. They would have to face the possibility that beyond the natural world there is a further reality which transcends science.

Facing that possibility is absolutely unacceptable, however. The reason why is the subject of the next two chapters.

---

[2] Cairns-Smith's answer is that he is inclined to the "majority prejudice," which is that the "exorcism [of supernatural forces] that Darwin initiated will continue right back to the origin of life."

Chapter Nine

The Rules of Science

In 1981, THE Arkansas state legislature passed a statute requiring "balanced treatment to creation-science and to evolution-science." Opponents sued in the local federal court to have the statute declared unconstitutional, and the stage was set for a very unequal contest.

The Arkansas statute was the work of unsophisticated activists who had no idea how to attract support from outside their own narrowly fundamentalist camp. As a result, they faced a powerful coalition of groups eager to defend both science and liberal religion against religious extremists. The coalition included not only the major associations of scientists and educators, but also the American Civil Liberties Union and an impressive array of individuals and organizations representing mainstream Christianity and Judaism.

The coalition also had the services of a first-class team of trial lawyers donated by one of America's biggest and best law firms. These specialists in "big-case" litigation knew how to select and prepare religious and scientific leaders to give expert testimony that would establish creation-science as an absurdity unworthy of serious consideration. Orthodox science won the case by a light-year.

Judge William Overton's decision distilled the testimony of the expert witnesses, especially the Darwinist philosopher of science Michael Ruse, and provided a definition of "science" that made it quite clear why there can be no such thing as "creation-science." Judge Overton began by defining science as whatever is "accepted by the scientific community," meaning of course the *official* scientific community. That in itself wasn't very informative, but the judge went on to specify five essential characteristics of science:

(1) It is guided by natural law;
(2) It has to be explanatory by reference to natural law;
(3) It is testable against the empirical world;
(4) Its conclusions are tentative—that is, not necessarily the final word; and
(5) It is falsifiable.

Creation-science does not meet these criteria, according to Judge Overton, because it appeals to the supernatural, and hence is not testable, falsifiable, or "explanatory by reference to natural law." As a typical illustration of the unscientific nature of creationist claims, the judge quoted the following statement by the creation-scientist Duane Gish:

> We do not know how God created, what processes He used, for God used processes which are not now operating anywhere in the natural universe. This is why we refer to divine creation as Special Creation. We cannot discover by scientific investigation anything about the creative processes used by God.

At the same time, Judge Overton indignantly denied the creationist claim that "belief in a creator and acceptance of the scientific theory of evolution are mutually exclusive," describing this opinion as "offensive to the religious views of many."

Philosophers of science have found much fault with Judge Overton's definition, and have hinted that Ruse and the other experts got away with a philosophical snow job. These critics pointed out that scientists are not in the least "tentative" about their basic commitments, including their commitment to evolution. In addition, scientists have often studied the effects of a phenomenon (such as gravity) which they could not explain by natural law. Finally, the critics observed that creation-science makes quite specific empirical claims (a young earth, a worldwide flood, special creation), which mainstream science has said are provably false. How can the same statements be both demonstrably false and unfalsifiable?

If the Ruse-Overton definition failed to satisfy the philosophers, however, it delighted the scientific establishment. The premier scientific journal *Science* was so enthusiastic that it reprinted the entire opinion as a major article. The opinion summed up the way many working scientists view their enterprise, which makes it a good starting point for discussing what science includes and excludes.

I am not interested in any claims that are based upon a literal reading of the Bible, nor do I understand the concept of creation as narrowly as Duane Gish does. If an omnipotent Creator exists He might have created things instantaneously in a single week or through gradual evolution over billions of years. He might have employed means wholly inaccessible to science, or mechanisms that are at least in part understandable through scientific investigation.

The essential point of creation has nothing to do with the timing or the mechanism the Creator chose to employ, but with the element of design or purpose. In the broadest sense, a "creationist" is simply a person who believes that the world (and especially mankind) was *designed*, and exists for a *purpose*. With the issue defined that way, the question becomes: Is mainstream science opposed to the possibility that the natural world was designed by a Creator for a purpose? If so, on what basis?

Judge Overton was persuaded that "creation" (in the general sense of design) is consistent with "evolution" in the scientific sense. In this he was mistaken, or rather, misled. When evolutionary biologists speak of "evolution," they are not referring to a process that either was or could have been guided by a supernatural Creator. They mean *naturalistic* evolution, a purely materialistic process that has no direction and reflects no conscious purpose. For example, here is how George Gaylord Simpson defined "the meaning of evolution":

Although many details remain to be worked out, it is already evident that all the objective phenomena of the history of life can be explained by purely naturalistic or, in a proper sense of the sometimes abused word, materialistic factors. They are readily explicable on the basis of differential reproduction in populations (the main factor in the modern conception of natural selection) and of the mainly random interplay of the known processes of heredity. . . . *Man is the result of a purposeless and natural process that did not have him in mind.* [Emphasis added.]

Because the scientific establishment has found it prudent to encourage a degree of confusion on this point, I should emphasize that Simpson's view was not some personal opinion extraneous to his scientific discipline. On the contrary, he was merely stating explicitly what Darwinists mean by "evolution." The same understanding is expressed in countless books and articles, and where it is not expressed it is pervasively implied. Make no mistake about it. In the Darwinist view, which is the official view of mainstream science, God had nothing to do with evolution.[1]

Theistic or "guided" evolution has to be excluded as a possibility because Darwinists identify science with a philosophical doctrine known as *naturalism*.[2] Naturalism assumes the entire realm of nature to be a closed system of material causes and effects, which cannot be influenced by anything from "outside." Naturalism does

---

[1] A second passage from Simpson's *The Meaning of Evolution* clarifies the relationship between naturalism and atheism. Scientific naturalists are not necessarily opposed to "the existence of God," provided that God is defined as an unreachable First Cause and not as a Creator who takes an active role in nature or human affairs. In Simpson's words:

There is neither need nor excuse for postulation of nonmaterial intervention in the origin of life, the rise of man, or any other part of the long history of the material cosmos. Yet the origin of that cosmos and the causal principles of its history remain unexplained and inaccessible to science. Here is hidden the First Cause sought by theology and philosophy. The First Cause is not known and I suspect it will never be known to living man. We may, if we are so inclined, worship it in our own ways, but we certainly do not comprehend it.

[2] A variety of terms have been used in the literature to designate the philosophical position I call scientific naturalism. For present purposes, the following terms may all be considered equivalent: scientific naturalism, evolutionary naturalism, scientific materialism, and scientism. All these terms imply that scientific investigation is either the exclusive path to knowledge or at least by far the most reliable path, and that only natural or material phenomena are real. In other words, what science can't study is effectively unreal.

not explicitly deny the mere existence of God, but it does deny that a supernatural being could in any way influence natural events, such as evolution, or communicate with natural creatures like ourselves. *Scientific* naturalism makes the same point by starting with the assumption that science, which studies only the natural, is our only reliable path to knowledge. A God who can never do anything that makes a difference, and of whom we can have no reliable knowledge, is of no importance to us.

Naturalism is not something about which Darwinists can afford to be tentative, because their science is based upon it. As we have seen, the positive evidence that Darwinian evolution either can produce or has produced important biological innovations is nonexistent. Darwinists know that the mutation-selection mechanism can produce wings, eyes, and brains not because the mechanism can be observed to do anything of the kind, but because their guiding philosophy assures them that no other power is available to do the job. The absence from the cosmos of any Creator is therefore the essential starting point for Darwinism.

The first two elements of Judge Overton's definition express the commitment of science to naturalism. The remaining three elements state its commitment to *empiricism*. A good empiricist insists that conclusions be supported by observation or experiment, and is willing to discard even the most cherished doctrines if they do not fit the evidence. Naturalism and empiricism are often erroneously assumed to be very nearly the same thing, but they are not. In the case of Darwinism, these two foundational principles of science are in conflict.

The conflict arises because creation by Darwinist evolution is hardly more observable than supernatural creation by God. Natural selection exists, to be sure, but no one has evidence that it can accomplish anything remotely resembling the creative acts that Darwinists attribute to it. The fossil record on the whole testifies that whatever "evolution" might have been, it was not the process of gradual change in continuous lineages that Darwinism implies. As an explanation for modifications in populations, Darwinism is an empirical doctrine. As an explanation for how complex organisms came into existence in the first place, it is pure philosophy.

If empiricism were the primary value at stake, Darwinism would long ago have been limited to microevolution, where it would have no important theological or philosophical implications. Such a limitation would not imply acceptance of creationism, even in the least restrictive definition of that term. What it *would* imply is that the scientific establishment after 1859 was carried away by enthusiasm, and thought it had proved an entire creation story when it had only filled in some minor details. If Darwinists accepted the primacy of empiricism, they could still hope eventually to find a naturalistic explanation for everything, but for now they would have to admit that they have made a big mistake.

That admission has not come, because empiricism is *not* the primary value at stake. The more important priority is to maintain the naturalistic worldview and with it the prestige of "science" as the source of all important knowledge. Without Darwinism, scientific naturalism would have no creation story. A retreat on a matter of this importance would be catastrophic for the Darwinist establishment, and it would open the door to all sorts of false prophets and mountebanks (at least as naturalists see them) who would try to fill the gap.

To prevent such a catastrophe, defenders of naturalism must enforce rules of procedure for science that preclude opposing points of view. With that accomplished, the next critical step is to treat "science" as equivalent to truth and non-science as equivalent to fantasy. The conclusions of science can then be misleadingly portrayed as refuting arguments that were in fact disqualified from consideration at the outset. As long as scientific naturalists make the rules, critics who demand positive evidence for Darwinism need not be taken seriously. They do not understand "how science works."

I am not implying that scientific naturalists do any of this with an intent to deceive. On the contrary, they are as a rule so steeped in naturalistic assumptions that they are blind to the arbitrary elements in their thinking. For example, examine carefully the following passage from *The Dreams of Reason*, a book about scientific reasoning, by Heinz Pagels:

So powerful is [the scientific-experimental] method that virtually everything scientists know about the natural world comes from it. What they find is that the architecture of the universe is indeed built according to invisible universal rules, what I call the cosmic code—

the building code of the Demiurge.[3] Examples of this universal building code are the quantum and relativity theory, the laws of chemical combination and molecular structure, the rules that govern protein synthesis and how organisms are made, to name but a few. Scientists in discovering this code are deciphering the Demiurge's hidden message, the tricks he used in creating the universe. No human mind could have arranged for any message so flawlessly coherent, so strangely imaginative, and sometimes downright bizarre. It must be the work of an Alien Intelligence!

. . . Whether God is the message, wrote the message, or whether it wrote itself is unimportant in our lives. We can safely drop the traditional idea of the Demiurge, for there is no scientific evidence for a Creator of the natural world, no evidence for a will or purpose that goes beyond the known laws of nature. Even the evidence of life on earth, which promoted the compelling "argument from design" for a Creator, can be accounted for by evolution. [Pagels refers his readers to books by Dawkins and Gould for the evidence.] So we have a message without a sender.

The first paragraph of that passage tells us that the presence of intelligent design in the cosmos is so obvious that even an atheist like Pagels cannot help noticing it, and rhapsodizing about it, dubbing the Creator "the Demiurge." The second paragraph offhandedly remarks that there is no scientific evidence for a Creator. What makes the passage a good illustration of the scientific naturalist mentality is that Pagels assumes all the critical points. What seemed to be evidence of a Creator turned out to be no evidence at all, because scientific evidence for something that goes beyond the laws of nature would be a contradiction in terms. On the other hand, evidence of "evolution" (which may mean no more than microevolution plus the existence of natural relationships) automatically excludes the possibility of design. Naturalistic philosophy controls his mind so completely that Pagels can stare straight at evidence of intelligent design, describe it as such, and still not see it.

The "will of the Creator" is a concept generally acknowledged to

---

[3] "Demiurge" is a term derived from Greek philosophy and the Gnostic heresy of early Christianity. The Gnostics considered matter to be evil and thought God would not have created it, and so they attributed the material world to the Demiurge, an inferior deity which they sometimes identified with the God of the Old Testament.

be outside the ken of natural science altogether. To a clear understanding, that means that science cannot tell us whether there is or is not a transcendent will or purpose that goes beyond the laws of nature. To a scientific naturalist, however, "outside of science" means outside of *reality*.

That is why scientific naturalists can in good conscience say at one moment that they do not deal with God or religion, and then in the next breath make sweeping pronouncements about the purposelessness of the cosmos. What other people understand as the limitations of science become twisted into limitations upon reality, because to scientific naturalists the notion that there could be a reality outside of science is literally unthinkable.

This way of thinking is encouraged by the way science employs paradigms as organizing concepts in guiding research. According to the famous model of Thomas Kuhn, the progress of science is much like Gould and Eldredge's theory of evolution by punctuated equilibrium. Periods of stasis, Kuhn's "normal science," are punctuated by revolutions in the form of "paradigm shifts," where one way of thinking about the subject is replaced by another. Like other philosophical theories, Kuhn's model has to be applied with caution. But whatever its limitations as a description of science generally, it provides an illuminating picture of the methodology of Darwinism.

The most important of Kuhn's concepts is the *paradigm*, which is not a mere theory or hypothesis but a way of looking at the world that is influenced by cultural prejudice as well as by scientific observation and experience. According to Kuhn, "An apparently arbitrary element, compounded of personal and historical accident, is always a formative ingredient of the beliefs espoused by a given scientific community at a given time." Scientists, like the rest of us, view reality through the mist of ideas and assumptions that make up the paradigm.

When a paradigm becomes established, it serves as the grand organizing principal for scientific research. This means that it defines the questions that need to be answered and the facts that need to be assembled. While the paradigm remains effectively unchallenged, "normal science" proceeds to work out its theoretical and practical implications and to solve the "puzzles" created by facts that do not seem to fit the paradigm's explanations. Science can make great progress during these periods, because scientists share a common understanding of what they are trying to do and how they should be trying to do it, and they are not distracted by uncertainty over fundamental assumptions. According to Kuhn:

> Closely examined, whether historically or in the contemporary labo-ratory, [normal science] seems an attempt to force nature into the preformed and relatively inflexible box that the paradigm supplies. No part of the aim of normal science is to call forth new sorts of phenomena; *indeed those that will not fit the box are often not seen at all.* Nor do scientists aim to invent new theories, and they are often intolerant of the theories invented by others. Instead, normal-scientific research is directed to the articulation of those phenomena and theories that the paradigm already supplies. [Emphasis added.]

Some puzzles prove recalcitrant to solution and gradually "anom-alies" build up. These do not threaten the dominance of the para-digm as long as research proceeds satisfactorily in other respects. Even a relatively inadequate paradigm can define a field of science and set an agenda for research, and it may take a long time for scientists to become convinced that some important problems will never be solved within the concepts of the existing paradigm. As Kuhn describes it, however, the intense commitment to the para-digm produces both the success of normal science and an inevitable crisis:

> Normal science, the activity in which most scientists inevitably spend almost all their time, is predicated on the assumption that the scien-tific community knows what the world is like. Much of the success of the enterprise derives from the community's willingness to defend that assumption, if necessary at considerable cost. Normal science, for example, often suppresses fundamental novelties because they are necessarily subversive of its basic commitments. Nevertheless, so long as those commitments retain an element of the arbitrary, the very nature of normal research ensures that novelty shall not be sup-pressed for very long.

Eventually, it becomes impossible to deny that there are problems which cannot be solved within the accepted way of looking at things. At this point a state of "crisis" is reached, and the field seems threatened by a pervasive confusion and chaos. The crisis is resolved by the emergence of a new paradigm, and normal science can proceed once again with confidence.

One influential definition of science which Kuhn's model challenged was the "falsifiability" criterion of the philosopher Karl Popper, which reappeared nonetheless as an element in Judge Overton's definition. Popper thought that a theory or hypothesis was scientific only to the extent that it was in principle capable of being shown to be false through empirical testing. The problem with this criterion is that it is impossible to test every important scientific proposition in isolation. Background assumptions have to be made so that detailed statements can be tested. The paradigm is made up of the background assumptions that define the current scientific worldview.

A paradigm is not merely a hypothesis, which can be discarded if it fails a single experimental test; it is a way of looking at the world, or some part of it, and scientists understand even the anomalies in its terms. According to Kuhn, anomalies by themselves never falsify a paradigm, because its defenders can resort to *ad hoc* hypotheses to accommodate any potentially disconfirming evidence. A paradigm rules until it is replaced with another paradigm, because "To reject one paradigm without substituting another is to reject science itself." The rule against "negative argumentation" which the National Academy of Sciences invoked in the Supreme Court case was an application of this logic.

When a new paradigm emerges it does more than explain the anomalies: it reorients the scientific perspective so strongly that the former anomalies may seem no longer to be mere facts but virtual tautologies, statements of situations that could not conceivably have been otherwise. It is therefore not as exceptional as it may have appeared that distinguished scientists have praised Darwin's theory as a profound tautology, or declared it to be a logically self-evident proposition requiring no empirical confirmation. A tautology or logical inevitability is precisely what the theory appears to them to be: it describes a situation that could not conceivably have been otherwise. From this perspective, "disconfirming" evidence is profoundly uninteresting.

Kuhn described experimental evidence showing that ordinary people tend to see what they have been trained to see, and fail to see what they know ought not to be present. The finest scientists are no exception; on the contrary, because they are dependent upon inferences and upon observations that are difficult to make, they are particularly prone to paradigm-influenced misperception.

Kuhn cited examples of visible celestial phenomena that were not "seen" until the new astronomical paradigm of Copernicus legitimated their existence. If Kuhn had chosen evolutionary biology as a case study, he would have risked being denounced as a creationist. As we saw in Chapter Four, the pervasive pattern of stasis in the fossil record long went unrecognized because to Darwinists it was not worth describing in print. The problem of tunnel vision is not something that can be expected to go away as science becomes more sophisticated. On the contrary, as essential funding is brought more and more under centralized governmental control, researchers have no alternative but to concentrate upon the agenda set by the paradigm.

A new paradigm does not merely propose different answers to the questions scientists have been asking, or explain the facts differently; it suggests entirely different questions and different factual possibilities. For this reason, opposing paradigms are to a certain extent "incommensurable," in the sense that their respective adherents find it difficult to communicate intelligibly with each other. Kuhn's insight in this respect is particularly true when the paradigm is not a specific scientific theory but a broad philosophical outlook.

To cite an example from my personal experience, it is pointless to try to engage a scientific naturalist in a discussion about whether the neo-Darwinist theory of evolution is *true*. The reply is likely to be that neo-Darwinism is the best scientific explanation we have, and that *means* it is our closest approximation to the truth. Naturalists will usually concede that any theory can be improved, and that our understanding of naturalistic evolution may one day be much greater than it is now. To question whether naturalistic evolution itself is "true," on the other hand, is to talk nonsense. Naturalistic evolution is the only conceivable explanation for life, and so the fact that life exists proves it to be true.

It is easy to see why scientific naturalism is an attractive philosophy for scientists. It gives science a virtual monopoly on the production of knowledge, and it assures scientists that no important questions are in principle beyond scientific investigation. The important question, however, is whether this philosophical viewpoint is merely an understandable professional prejudice or whether it is *the* objectively valid way of understanding the world. That is the real issue behind the push to make naturalistic evolution a fundamental tenet of society, to which everyone must be converted.

If scientific naturalism is to occupy a dominant cultural position, it must do more than provide information about the physical universe. It must draw out the spiritual and ethical implications of its creation story. In short, evolution must become a religion. We shall see in the following chapters how this has been accomplished.

Chapter Ten

Darwinist Religion

THE PREFACE TO the 1984 pamphlet *Science and Creationism: A View From the National Academy of Sciences*, signed by the Academy's president, Frank Press, assured the nation that it is "false . . . to think that the theory of evolution represents an irreconcilable conflict between religion and science." Dr. Press explained:

> A great many religious leaders accept evolution on scientific grounds without relinquishing their belief in religious principles. As stated in a resolution by the Council of the National Academy of Sciences in 1981, however, "Religion and science are separate and mutually exclusive realms of human thought whose presentation in the same context leads to misunderstanding of both scientific theory and religious belief."

The Academy's concern was only to justify its opposition to creation-science, and it did not feel obliged to explain what "religion" might be, or under what circumstances the religious realm

123

might be entitled to protection from incursions by science. Stephen Jay Gould had somewhat more to say on this subject, however, in his rebuttal to Irving Kristol's charge that neo-Darwinism as currently taught incorporates "an ideological bias against religious belief." Gould responded that most scientists show no hostility to religion, because their subject "doesn't intersect the concerns of theology."

> Science can no more answer the question of how we ought to live than religion can decree the age of the earth. Honorable and discerning scientists (most of us, I trust) have always understood that the limits to what science can answer also describe the power of its methods in their proper domain. Darwin himself exclaimed that science couldn't touch the problem of evil and similar moral conundrums: "A dog might as well speculate on the mind of Newton. Let each man hope and believe what he can."

The Gould-Darwin disclaimer contains an important ambiguity. If science can tell us nothing about how we ought to live, does this mean that knowledge about this subject can be obtained through religion, or does it mean that we can know no more about good and evil than a dog knows about the mind of Newton? Each man may hope and believe as he can, but there are some who would say that hopes and beliefs are mere subjective expressions of feeling, little more than sentimental nonsense, unless they rest upon the firm foundation of scientific knowledge.

One Darwinist who says exactly this is Cornell University Professor William Provine, a leading historian of science. Provine insists that the conflict between science and religion is inescapable, to the extent that persons who manage to retain religious beliefs while accepting evolutionary biology "have to check [their] brains at the church-house door." Specifically:

> Modern science directly implies that the world is organized strictly in accordance with mechanistic principles. There are no purposive principles whatsoever in nature. There are no gods and no designing forces that are rationally detectable. . . .
>
> Second, modern science directly implies that there are no inherent moral or ethical laws, no absolute guiding principles for human society.
>
> Third, human beings are marvelously complex machines. The

individual human becomes an ethical person by means of two primary mechanisms: heredity and environmental influences. That is all there is.

Fourth, we must conclude that when we die, we die and that is the end of us. . . .

Finally, free will as it is traditionally conceived—the freedom to make uncoerced and unpredictable choices among alternative possible courses of action—simply does not exist. . . . There is no way that the evolutionary process as currently conceived can produce a being that is truly free to make choices.

Gould had assured Kristol that among evolutionary biologists there is "an entire spectrum of religious attitudes—from devout daily prayer and worship to resolute atheism." I have myself noticed a great deal more of the latter than the former, and Provine agrees with me. He reports that most evolutionary biologists are atheists, "and many have been driven there by their understanding of the evolutionary process and other science." The few who see no conflict between their biology and their religion "are either obtuse or compartmentalized in their thinking, or are effective atheists without realizing it." Scientific organizations hide the conflict for fear of jeopardizing the funding for scientific research, or because they feel that religion plays a useful role in moral education. According to Provine, who had the Academy's 1984 statement specifically in mind, "These rationalizations are politic but intellectually dishonest."

It is not difficult to reconcile all these statements, once we untangle the confusing terminology. The Academy is literally correct that there is no incompatibility between "evolution" and "religion." When these terms are not defined specifically, neither has enough content to be incompatible with anything else. There is not even any conflict between evolution and *theistic* religion. God might very well have "created" by gradually developing one kind of creature out of other kinds. Evolution of that sort is not what the scientists have in mind, but they have nothing to gain from making this clear to the public.

Gould's remark is similarly misleading. Most scientific naturalists accept what is called the "fact-value distinction," and do not claim that a scientific description of what "is" can lead directly to a theory of what we "ought" to do. On the other hand, they do not consider all statements about ethics to be equally rational. A rational person starts with what is known and real rather than what is unknown and unreal. As George Gaylord Simpson explained the matter:

> Of course there are some beliefs still current, labelled as religious and involved in religious emotions, that are flatly incompatible with evolution and therefore are intellectually untenable in spite of their emotional appeal. Nevertheless, I take it as now self-evident, requiring no further special discussion, that evolution and *true* religion are compatible.

A scientific doctrine that sets the boundary between true and false religion is certainly not "anti-religious," but it directly contradicts the Academy's assurance that religion and science are separate and mutually exclusive realms of human thought.

Scientific naturalists do not see a contradiction, because they never meant that the realms of science and religion are of equal dignity and importance. Science for them is the realm of objective knowledge; religion is a matter of subjective belief. The two should not conflict because a rational person always prefers objective knowledge to subjective belief, when the former is available. Religions which are based on intellectually untenable ideas (such as that there is a Creator who has somehow communicated His will to humans) are in the realm of fantasy. Naturalistic religion, which looks to science for its picture of reality, is a way of harnessing irrational forces for rational purposes. It may perform useful service by recruiting support for scientific programs in areas like environmental protection and medical research.

The American Scientific Affiliation (ASA) incurred the wrath of Darwinists for mixing the wrong kind of religion with science. The ASA's membership is made up of science teachers and others who identify themselves as evangelical Christians committed both to Jesus Christ and to a scientific understanding of the natural world. The fundamentalist creation-scientists split from the ASA years ago in disgust at its members' willingness to accept not only the geological evidence that the earth is very old, but also the theory of biological evolution.

The ASA leadership has generally embraced "compatibilism" (the doctrine that science and religion do not conflict because they occupy separate realms) and "theistic evolution." Theistic evolution is not easy to define, but it involves making an effort to maintain that the natural world is God-governed while avoiding disagreement with the Darwinist establishment on scientific matters. Because the Darwinists have become increasingly explicit about the religious and philosophical implications of their system, this strategy led the theism in the ASA's evolution to come under ever greater pressure.

Compatibilism has its limits, however, and some ASA leaders were prodded into action by the strong naturalistic bias of the National Academy's 1984 pamphlet, which tried to give the public the impression that science has all the major problems of evolution well in hand. With foundation support, the ASA produced its own 48-page illustrated booklet, titled *Teaching Science in a Climate of Controversy: A View from the American Scientific Affiliation*, and mailed it to thousands of school teachers. The general tenor of the booklet was to encourage open-mindedness, especially on such "open questions" as whether life really arose by chance, how the first animals could have evolved in the Cambrian explosion, and how human intelligence and upright posture evolved.[1]

The ASA members who wrote *Teaching Science* naively expected that most scientists would welcome their contribution as a corrective to the overconfidence that evolutionary science tends to project when it is trying to persuade the public not to entertain any doubts. The official scientific organizations, however, are at war with cre-

---

[1] The following paragraphs reflect the general theme of *Teaching Science*:

Many aspects of evolution are currently being studied by scientists who hold varying degrees of belief or disbelief in God. No matter how those investigations turn out, most scientists agree that a 'creation science' based on an earth only a few thousand years old provides no theoretical basis sound enough to serve as a reasonable alternative.

Clearly, it is difficult to teach evolution—or even to avoid teaching it—without stepping into a controversy loaded with all kinds of implications: scientific, religious, philosophical, educational, political and legal. Dogmatists at either extreme who insist that theirs is the only tenable position tend to make both sides seem unattractive.

Many intelligent people, however, who accept the evidence for an earth billions of years old and recognize that life-forms have changed drastically over much of that time, also take the Bible seriously and worship God as their Creator. Some (but not all) who affirm creation on religious grounds are able to envision *macro*-evolution as a possible explanation of how God has created new life-forms.

In other words, a broad middle ground exists in which creation and evolution are not seen as antagonists.

ationism, and their policy is to demand unconditional surrender. Persons who claim to be scientists, but who try to convince school teachers that there are "open questions" about the naturalistic understanding of the world, are traitors in that war.

Retribution quickly followed. A California "science consultant" named William Bennetta, who makes a career of pursuing creationists, organized a posse of scientific heavyweights to condemn the ASA's pamphlet as "an attempt to replace science with a system of pseudoscience devoted to confirming Biblical narratives." A journal called *The Science Teacher* published a collection of essays edited by Bennetta, titled "Scientists Decry a Slick New Packaging of Creationism." Nine prominent scientists, including Gould, Futuyma, Eldredge, and Sarich, contributed heavy-handed condemnations of *Teaching Science*. The pervasive message was that the ASA is a deceitful creationist front which disguises its Biblical literalist agenda under a pretence of scientific objectivity.

The accusations bewildered the authors of *Teaching Science*, and were so far off the mark that persons familiar with the ASA might easily have mistaken them for intentional misrepresentations. It would be a mistake to infer any intent to deceive, however, because really zealous scientific naturalists do not recognize subtle distinctions among theists. To the zealots, people who say they believe in God are either harmless sentimentalists who add some vague God-talk to a basically naturalistic worldview, or they are creationists. In either case they are fools, but in the latter case they are also a menace.

From a zealot's viewpoint, the ASA writers had provided ample evidence of a creationist purpose. Why would they harp on "open questions" except to imply that God might have taken a hand in the appearance of new forms? That suggestion is creationism by definition, and the ASA admits to being an organization of Christians who accept the authority of the Bible. Their true reason for rejecting scientific evolution must therefore be that it contradicts the Biblical narrative. What other reason could they have?

Mixing religion with science is obnoxious to Darwinists only when it is the wrong religion that is being mixed. To prove the point, we may cite two of the most important founders of the modern synthesis, Theodosius Dobzhansky and Julian Huxley. Julian Huxley's religion of "evolutionary humanism" offered humanity the "sacred duty" and the "glorious opportunity" of seeking "to promote the maximum fulfillment of the evolutionary process on the earth." That did not mean merely working to ensure that the organisms that have the most offspring continue to have the most offspring, but rather promoting the "fullest realization" of mankind's "inherent possibilities." Inspired by the same vision, the American philosopher and educational reformer John Dewey launched a movement in 1933 for "religious humanism," whose Manifesto reflected the assumption current among scientific naturalists at the time that the final demise of theistic religion would usher in a new era of scientific progress and social cooperation for mankind. Soon thereafter, Hitler and Stalin provided a stunning realization of some of mankind's inherent possibilities. Dewey's successors admitted in 1973 that a new Manifesto was needed because the events of the previous forty years had made the original statement "seem far too optimistic."

The revised Manifesto makes some unenthusiastic concessions to reality, such as that "Science has sometimes brought evil instead of good," and "Traditional religions are surely not the only obstacle to human progress." The overall message is as before. It is that salvation comes through science:

> Using technology wisely, we can control our environment, conquer poverty, markedly reduce disease, extend our life-span, significantly modify our behavior, alter the course of human evolution and cultural development, unlock vast new powers, and provide humankind with unparalleled opportunity for achieving an abundant and meaningful life.

The scientist-philosopher who went farther than anybody else in drawing a message of cosmic optimism from evolution was Pierre Teilhard de Chardin, the unorthodox Jesuit paleontologist who played an important role in the Piltdown and Peking Man discoveries. Teilhard aimed to bring Christianity up to date by founding it squarely upon the rock of evolution rather that upon certain events alleged to have occurred in Palestine nearly two thousand years ago. The more rigorously materialistic Darwinists dismissed Teilhard's philosophy as pretentious claptrap, but it had a strong appeal to those of a more spiritual cast of mind, such as Theodosius Dobzhansky.

In his reply to Irving Kristol, Gould cited Dobzhansky, "the greatest evolutionist of our century and a lifelong Russian Orthodox," to illustrate the compatibility of evolution and religion. For Dobzhansky the two were a good deal more than compatible, for he wrote in his book *Mankind Evolving* that Darwin had healed "the wound inflicted by Copernicus and Galileo." This wound was the discovery that the earth, and therefore man, is not the physical center of the universe. Darwinism had healed it by placing mankind at the *spiritual* center of the universe, because man now understands evolution and has the potential capacity to take control of it. Dobzhansky exulted that "Evolution need no longer be a destiny imposed from without; it may conceivably be controlled by man, in accordance with his wisdom and his values." For further detail he referred his readers to the following quotations, which encapsulate Teilhard's "inspiring vision":

> Is evolution a theory, a system, or a hypothesis? It is much more—it is a general postulate to which all theories, all hypotheses, all systems must henceforth bow and which they must satisfy in order to be thinkable and true. Evolution is a light which illuminates all facts, a trajectory which all lines of thought must follow—this is what evolution is.

Evolution is, in short, the God we must worship. It is taking us to heaven, "The Point Omega" in Teilhard's jargon, which is:

> a harmonized collectivity of consciousness, equivalent to a kind of superconsciousness. The earth is covering itself not only by myriads of thinking units, but by a single continuum of thought, and finally forming a functionally single Unit of Thought of planetary dimensions. The plurality of individual thoughts combine and mutually reinforce each other in a single act of unanimous Thought. . . . In the dimension of Thought, like in the dimension of Time and Space, can the Universe reach consummation in anything but the Measureless?

The naive optimism of these attempts to fashion a scientific religion survives in the contemporary "New Age" movement, but the trend among Darwinists today is to take a more somber view of humanity's prospects. Writing in 1989, Maitland Edey and Donald Johanson speculate that Homo sapiens may be about to make itself extinct, as a result of nuclear war or ecological catastrophe. This depressing situation is the result of a runaway technology that produces enormous quantities of toxic waste, destroys the jungle and the ozone layer, and permits unrestrained population growth. We are unable to deal intelligently with these problems because "in our guts we are passionate stone age people" who are capable of creating technology but not controlling it. Edey and Johanson think that science is about to develop the technical capacity to design "better people" through genetic engineering. If humanity is to avoid extinction, it must summon the political will to take control of evolution, and make it in the future a matter of human choice rather than blind selection.

The continual efforts to base a religion or ethical system upon evolution are not an aberration, and practically all the most prominent Darwinist writers have tried their hand at it. Darwinist evolution is an imaginative story about who we are and where we came from, which is to say it is a creation myth. As such it is an obvious starting point for speculation about how we ought to live and what we ought to value. A creationist appropriately starts with God's creation and God's will for man. A scientific naturalist just as appropriately starts with evolution and with man as a product of nature.

In its mythological dimension, Darwinism is the story of humanity's liberation from the delusion that its destiny is controlled by a power higher than itself. Lacking scientific knowledge, humans at first attribute natural events like weather and disease to supernatural beings. As they learn to predict or control natural forces they put aside the lesser spirits, but a more highly evolved religion retains the notion of a rational Creator who rules the universe.

At last the greatest scientific discovery of all is made, and modern humans learn that they are the products of a blind natural process that has no goal and cares nothing for them. The resulting "death of God" is experienced by some as a profound loss, and by others as a liberation. But liberation to what? If blind nature has somehow produced a human species with the capacity to rule earth wisely, and if this capacity has previously been invisible only because it was smothered by superstition, then the prospects for human freedom and happiness are unbounded. That was the message of the Humanist Manifesto of 1933.

Another possibility is that purposeless nature has produced a world ruled by irrational forces, where might makes right and human freedom is an illusion. In that case the right to rule belongs to whoever can control the use of science. It would be illogical for the rulers to worry overmuch about what people say they *want*, because science teaches them that wants are the product of irrational forces. In principle, people can be made to want something better. It is no kindness to leave them as they are, because passionate stone age people can do nothing but destroy themselves when they have the power of scientific technology at their command.

Whether a Darwinist takes the optimistic or the pessimistic view, it is imperative that the public be taught to understand the world as scientific naturalists understand it. Citizens must learn to look to science as the only reliable source of knowledge, and the only power capable of bettering (or even preserving) the human condition. That implies, as we shall see, a program of indoctrination in the name of public education.

Chapter Eleven

Darwinist Education

THE BRITISH MUSEUM of Natural History, located in a magnificent Victorian building in greater London's South Kensington district, celebrated its centennial in 1981 by opening a new exhibition on Darwin's theory. One of the first things a visitor encountered upon entering the exhibit was a sign which read as follows:

> Have you ever wondered why there are so many different kinds of living things?
>
> One idea is that all the living things we see today have EVOLVED from a distant ancestor by a process of gradual change.
>
> How could evolution have occurred? How could one species change into another?
>
> The exhibition in this hall looks at one possible explanation—the explanation of Charles Darwin.

An adjacent poster included the statement that "Another view is that God created all living things perfect and unchanging." A brochure asserted that "the concept of evolution by natural selection is not, strictly speaking, scientific," because it has been established by logical deduction rather than empirical demonstration. The brochure observed that "if the theory of evolution is true," it provides an explanation for the "groups-within-groups" arrangement of nature described by the taxonomists. The general tenor of the exhibit was that Darwinism is an important theory but not something which it is unreasonable to doubt.

Prominent scientists reacted furiously to these relativistic expressions. The forum for the controversy was the editorial and correspondence pages of the leading British science journal, *Nature*. L. B. Halstead, a neo-Darwinist stalwart, began things with a letter that attacked not only the Darwin exhibit but also new exhibits at the Museum on dinosaurs and human evolution. What was wrong with all these exhibits, according to Halstead, is that they employed a system of classification known as *cladism*, which assumes that no species can be identified as the ancestor of any other species.[1] He described the cladistic literature as full of "abuse of Ernst Mayr and George Gaylord Simpson, and indeed of Charles Darwin himself," because these great men had adhered firmly to "the idea that the processes that can be observed in the present day, when extrapolated into the past, are sufficient to explain the changes observed in the fossil record."

Halstead charged that some of the exhibits could be interpreted as attacking not only Darwinism, but evolution itself. For example, the exhibit on "Man's Place in Evolution" specifically denied that *Homo erectus* was a direct ancestor of *Homo sapiens*, so that "What the creationists have insisted on for years is now being openly advertised by the Natural History Museum."

It was not creationists that Halstead blamed for these transgres-

---

[1] Cladism has taken the science of biological classification by storm in recent years, and is now pervasively employed in museum exhibits and textbooks. For present purposes, the important point is that "cladograms" show relationships among living and fossil species, but never ancestral relationships. If two species (like chimp and man) are thought to resemble each other more closely than either resembles any third species, then the two are placed adjacent to each other in a cladogram. The hypothetical common ancestor that is supposed to be responsible for the relationship is never identified. Some Darwinists of the old school think that cladism predisposes the mind to think of evolution as a process of sudden branching rather than Darwinist gradualism, and a few cladists have said that, as far as their work is concerned, the hypothesis of common ancestry might as well be abandoned.

sions, however, but Marxists. Marxists tend to prefer a model of evolutionary change that proceeds by rapid bursts rather than by constant gradualism, because it fits with their view that social change occurs by a revolutionary leap from one kind of state to another. Darwin's gradualism, on the other hand, has unmistakable similarities to the model of step-by-step societal improvement through free economic competition and democratic reform that was so widely accepted in Victorian England. Halstead presented no concrete evidence of any Marxist motivation among the Museum's scientists, but he asserted that the Museum was "either unwittingly or willingly" giving support to Marxist theory by casting doubt upon Darwinist gradualism.[2]

The charge of political motivation was good entertainment, but the substantial issue was that the Museum's staff was "going public" with doubts about neo-Darwinism and even the existence of fossil ancestors—doubts that had previously been expressed only in professional circles. Specifically, some of the exhibits were suggesting that the orthodox theory finds its support in a certain kind of logic rather than in the scientific evidence. A report in *Nature* quoted what one of the Museum's senior scientists was telling the public in a film lecture:

> The survival of the fittest is an empty phrase; it is a play on words. For this reason, many critics feel that not only is the idea of evolution unscientific, but the idea of natural selection also. There's no point in asking whether or not we should believe in the idea of natural selection, because it is the inevitable logical consequence of a set of premises. . . .
>
> The idea of evolution by natural selection is a matter of logic, not science, and it follows that the concept of evolution by natural selection is not, strictly speaking, scientific. . . .
>
> If we accept that evolution *has* taken place, though obviously we must keep an open mind on it. . . .

---

[2] Although Halstead's charge was groundless, it is a fact that political ideology and biological ideology are often closely related. Prominent Darwinists such as Harvard's Richard Lewontin and Stephen Jay Gould have proudly claimed Marxist inspiration for their biological theories. Darwinists of the right have frequently related their biological theories to notions of economic or racial competition. At a scientific meeting in East Germany in 1981, the Darwinist philosopher of science Michael Ruse observed (with approval) that "Biology drips with as many wishes/wants/desires/urges, as many exhortations towards right actions, as a sermon by Luther or Wesley."

> We can't prove that the idea is true, only that it has not yet been proved false. It may one day be replaced by a better theory, but until then. . . .

The reporter commented: "If this is the voice of our friends and supporters, then Creation protect us from our enemies."

An editorial in *Nature* titled "Darwin's Death in South Kensington" hammered the offenders with rhetorical questions:

> Can it be that the managers of the museum which is the nearest thing to a citadel of Darwinism have lost their nerve, not to mention their good sense? . . . Nobody disputes that, in the public presentation of science, it is proper whenever appropriate to say that disputed matters are in doubt. But is the theory of evolution still an open question among serious biologists? And if not, what purpose, except general confusion, can be served by these weasel words?

The editorial speculated that the exhibition must have been designed by someone not in close contact with the museum's scientific staff, because most of those distinguished biologists "would rather lose their right hands than begin a sentence with the phrase 'If the theory of evolution is true'." This provoked an indignant response from 22 of the distinguished biologists, who were "astonished" that *Nature* would "advocate that theory be presented as fact." The biologists wrote that "we have no absolute proof of the theory of evolution," although we do have "overwhelming circumstantial evidence in favor of it and as yet no better alternative." They concluded, perhaps naively, that "the theory of evolution would be abandoned tomorrow if a better theory appeared."

The exchange of letters and editorial comments continued for months. The editors of *Nature* belatedly discovered that Darwinism is more controversial among scientists than they had realized, and they tried to take a more moderate line in a leading article on the boundaries of legitimate doubt. This effort—with the provocative title "How True is the Theory of Evolution?"—contributed to the general confusion by making concessions that must have been more alarming to the Darwinists than the exhibits at the Museum. The editors interpreted Karl Popper as having said that Darwinism is both metaphysical and unfalsifiable, unwisely conceded that this characterization is "technically correct," and then lamely responded that "the theory of evolution is not entirely without empirical support," and "metaphysical theories are not necessarily bad theories."

The rambling essay went on to acknowledge that "large sections of the general public are skeptical about Darwinism," and urged the Museum to challenge these skeptics, by throwing light upon the disputed issues. The skeptics were divided into two categories: "While some who doubt Darwinism do so on respectable grounds, others claim that the course of events may be determined by literally supernatural influences. Theories of that type are not even metaphysical—they are simply unscientific." The article ended by urging that "agnosticism" (about the absolute truth of scientific theories) not be "carried too far," to avoid demoralizing scientists. Although conceding that prejudice was in general to be avoided, *Nature* insisted that "one prejudice is allowable, even necessary—the preconception that theories can be constructed to account for all observable phenomena."

The *Nature* editorial not only implied that Darwinism is a metaphysical system sustained partly by faith, but courted outright disaster by encouraging the Museum to educate the public on the evidentiary problems that cause some people to become skeptical about Darwinism. Things could hardly be left at that point, and a few weeks later *Nature* published another article which tried to clean up the mess. It asserted that although "no biologist can deny the possibility that God created man, few would doubt that, if he did so, the mechanism that Darwin discerned was the one that He chose to use."[3] The Museum's duty was not to pander to doubters, but to make the case for evolutionism:

> In the face of the organized pressures of religious and mystical sects, evolutionists need some organization to represent their views, no less fervently held, as cogently as possible. Not that it should descend to the half-truths and doubletalk of political propaganda. But it should suit the terms of its message to those who will listen to it,

---

[3] Presumably the mechanism this writer had in mind was natural selection. The Darwin who wrote *The Descent of Man* was disenchanted with natural selection, however, half-apologized for giving it too much importance in *The Origin of Species*, and relied largely on sexual selection (and other vague mechanisms that would have little support among neo-Darwinists today) to explain the origin of human features.

rather than blunting its edge with the hair-splitting logic-chopping of the philosophy of science.

The cladists also scored some points in the debate. Particularly biting was the letter from Gareth Nelson:

> To the dismay, sometimes acute, of the more clerically minded members of this profession, cladistics treats fossils in a secular fashion—not as revelation but as some among many other biological specimens subject to interpretation that is apt, indeed expected, to be diverse, especially with respect to details.... As reasonable as this treatment might seem to the outsider, the emotional effect within such a paleontologist involuntarily confronted with cladistics (as I have witnessed on more occasions than I care to remember) is not unlike that experienced by a fundamentalist minister who has forced upon him uninvited the notion that the Bible is just one book among many. Suffice it to say that more than one kind of church has been built upon rock.

The view prevailed, however, that it would only mislead museum-goers to be presented with the notion that *The Origin of Species* is just one book among many. Anthony Flew, a philosopher renowned for defending Darwinism, atheism, and clear thinking, subsequently explained the whole episode as a breach of trust by "civil servants" (i.e. the Museum's scientists) who had a duty to present the established truth rather than to confuse the public with unorthodox opinions. He denounced these upstarts for their "abuse of the resources of a state institution to try to put [their pet theory, cladism] across to all the innocent and predominantly youthful laypersons who throng these public galleries, as if it were already part of the established consensus among all those best qualified to judge."

Flew reported that "the offending material has since, apparently and none too soon, been withdrawn." As this comment implied, the Museum had surrendered to the pressure. The Museum's spokesman explained (in a letter to *Nature*) that the staff's attempt to avoid dogmatism in its presentation of Darwinism had unfortunately given "an impression other than that intended." The film loop that had called survival of the fittest an empty phrase had been removed at once, and a more general cleanup of the exhibitions would follow.

When I visited the Museum in 1987, the exhibits contained nothing to alert the casual observer to the fact that there is anything controversial about Darwin's theory. For example, the infamous "one possible explanation" sign at the entrance to the Darwin exhibit had been replaced with the following reassuring message:

> When we compare ourselves with our fossil relatives, we find evidence that man has evolved.
> Darwin's work gave strong support to the view that all living things have developed into the forms we see today by a process of gradual change over very long periods of time.
> This is what we mean by *evolution*.
> Many people find that the theory of evolution does not conflict with their religious beliefs.

The "weasel words" in the original exhibit had hinted broadly that there were grounds for doubt about Darwinism, but had given no clear indication of precisely what the grounds for doubt might be. As the Museum's spokesman explained in an interview, the exhibits did not refer to such problems as the lack of transitionals in the fossil record, the sudden explosion of complex life forms at the beginning of the Cambrian age, the difficulty of explaining the origin of the genetic code, the limits to change shown by breeding experiments, the "hopeful monster" controversy, the punctuated equilibrium controversy, or the importance of catastrophic extinctions. From the point of view of an informed critic, even the original exhibition was more a coverup than a candid disclosure of Darwinism's difficulties. The spokesman pointed out that the Museum had nonetheless come a long way since the previous exhibit on evolution twenty years before, when the Director (Sir Gavin de Beer) "wrote a handbook in which it was said that these days, evolution is accepted as a fact, and natural selection is the mechanism for it, full stop. As far as he was concerned, the interesting conceptual bit of it was completely wrapped up, there was nothing left to think about."

The battle at the British Natural History Museum showed that creationists are not necessarily responsible for the fact that educators tend to stick to generalities when presenting the evidence for evolution to young people. Darwinists are very resentful if their theory is presented to the impressionable in a manner likely to encourage doubts. An explanation of the punctuated equilibrium controversy, for example, is bound to give skeptics the impression that Darwinists are making lame excuses for their inability to find supporting fossil evidence for their claims about macroevolution. No matter how earnestly the experts insist that they are only arguing about the *tempo* of gradualist evolution, and not about whether it ever happened, a few bright teenagers are likely to think that perhaps the evidence is missing because the step-by-step transitions never occurred. To Darwinists, teaching about evolution does not mean encouraging immature minds—or mature ones, for that matter—to think about unacceptable possibilities.

CALIFORNIA IS a state with a diverse population that includes many creationists and also a large and assertive scientific community. In the early 1970s, creationists persuaded the State Board of Education to adopt an "Antidogmatism Policy," but, more recently, science educators have counterattacked. They pressed the State Board of Education to enact clear rules mandating the teaching of evolution as Darwinists understand it.

After much debate the Board adopted a *Policy Statement* on the Teaching of Science in early 1989. Although the whole point of the new policy is to encourage more thorough coverage of evolution in classrooms and textbooks, the *Policy Statement* itself does not refer explicitly to evolution. The educators preferred to make a more general statement about "science" because they did not want to concede that evolution is an exceptional case which involves religious or philosophical questions distinct from those present in other areas of science.

On its face, the *Policy Statement* is reasonable and broad-minded. It begins by saying that science is concerned with observable facts and testable hypotheses about the natural world, and not with divine creation, ultimate purposes, or ultimate causes. These non-scientific subjects are relegated to the literature and social studies curricula. The *Policy Statement* emphasizes that neither science nor anything else should be taught dogmatically, because "Compelling beliefs is inconsistent with the goal of education," which is to encourage understanding. The *Policy Statement* even repeats this important distinction between believing and understanding: "To be fully informed citizens, students do not have to accept everything that is taught in the natural sciences curriculum, but they do have to understand the major strands of scientific thought, including its methods, facts, hypotheses, theories, and laws."

The *Policy Statement* goes on to explain that scientific facts, theories, and hypotheses are subject to testing and rejection; this feature distinguishes them from beliefs and dogmas, which do not meet the criterion of testability and are therefore inappropriate for consideration in science classes. Science teachers are professionally obligated to stick to science, and should respectfully encourage students to discuss matters outside the domain of science with their families and clergy.

A person unaware of the nuances of the knowledge-belief distinction might imagine that the *Policy Statement* protects the right of creationist students to question the truth of evolution, provided they "understand" the subject. That would be a misunderstanding, however, because from a Darwinist perspective it is no more possible to understand evolution and honestly disbelieve it than it is to understand arithmetic and think that four times two is seven. To Darwinists, fully naturalistic evolution is a fact to be learned, not an opinion to be questioned. A student may silently disbelieve, but neither students nor teachers may discuss the grounds for disbelief in class, where other students might be infected.

The purpose of the *Policy Statement* is not to protect dissent, but to establish a philosophical justification for teaching naturalistic evolution as "fact" in an educational system that is at least nominally opposed to dogmatism. The justification is that science is a world apart because of the exceptional reliability of its methods. Scientific facts and theories are subject to continual testing, whereas philosophical and religious beliefs "are based, at least in part, on faith, and are not subject to scientific test and refutation." Although compelling *beliefs* is inconsistent with the goal of education, compelling *knowledge* is what education is all about. Those who understand the code words know that all these generalities are meant to establish a single specific point: naturalistic evolution belongs in the category of knowledge, not belief, and so resistance to it stems from ignorance, which education rightly aims to eliminate.

The *Policy Statement* was followed by a curriculum guide called the *Science Framework*, which tells textbook publishers what approach to take if they want their books to be acceptable in the huge California market. The *Framework* pays lip service to the principle that teaching should be nondogmatic, but it also conveys a clear message that the purpose of instruction in evolution is to persuade students to believe in the orthodox theory. The major areas of difficulty are ignored or minimized. Teachers are exhorted to reassure students that science is a reliable and self-correcting enterprise, that allegedly scientific objections to accepted doctrines have been considered and rejected by the scientific community, and that evolution is "scientifically accepted fact."

The language in which all this is done seems calculated more to conceal information than to reveal it. For example, instead of acknowledging that science cannot demonstrate how complex adaptive structures can arise by random mutation and selection, the *Framework* provides a pointless distinction between "natural selection" and "adaptation."

> Natural selection and adaptation are different concepts. Natural selection refers to the process by which organisms whose biological characteristics better fit them to their environments are better represented in future generations. . . . Adaptation is the process by which organisms respond to the challenges of their environments, through natural selection with changes and variations in their form and behavior.

The inability of paleontologists to identify specific fossil ancestors for any of the major groups is addressed obliquely in one sentence: "Discovering evolutionary relationships is less a search for ancestors than for groups that are most closely related to each other." The notorious controversies over the pace of macroevolution are papered over with the observation that gradualism is the rule except when it is not the rule.

> Although most changes in organisms occur in small steps over a long period of time, some major biological changes have taken place during relatively short intervals and at certain points in the earth's history. These include the evolution, diversification, and extinction of much fossil life.

Finally, the *Framework* includes a table to illustrate the extreme regularity in cytochrome c sequence divergences. This so-called "molecular clock" phenomenon contradicted expectations based on the theory of natural selection, and required the invention of the neutral theory of molecular evolution. The *Framework* comments that the table "shows how regular has been the rate of molecular evolution in these amino acid sequence changes. Its results are exactly what would be expected and predicted by evolutionary theory."[4]

In its introductory section, the *Framework*'s authors extol science as "a limitless voyage of joyous exploration," and stress the importance of inspiring students with the excitement of the scientific enterprise. That sense of excitement is not supposed to extend to fundamental questions about evolution, however. Students are encouraged to think about careers in biotechnology, but solving the mystery of evolution is out of the question because Darwinists have to insist that there is no mystery. The "interesting conceptual bit" has been settled, and only the details remain to be filled in.

The *Framework*'s most constructive recommendation is that teachers and textbook writers should avoid terminology that implies that scientific judgments are a matter of subjective preference or vote-counting.

> Students should never be told that "many scientists" think this or that. Science is not decided by vote, but by evidence. Nor should students be told that "scientists believe." Science is not a matter of belief; rather, it is a matter of evidence that can be subjected to the tests of observation and objective reasoning. . . . Show students that nothing in science is decided just because someone important says it is so (authority) or because that is the way it has always been done (tradition).

The *Framework* immediately contradicts that message, however, by defining "evolution" only vaguely, as "change through time." A vaguely defined concept cannot be tested by observation and objective reasoning. The *Framework* then urges us to believe in this vague concept because so many scientists do: "It is an accepted scientific explanation and therefore no more controversial in scientific circles than the theories of gravitation and electron flow." An appeal to

---

[4] The cytochrome c table caused embarrassment to the Framework's authors when it was discovered to contain typographical errors identical to those in a similar table printed in a creationist textbook titled *Of Pandas and People*. Confronted with the evidence, the consultant responsible for the evolutionary biology sections of the Framework admitted that he had copied the table from the creationist book, reversing the order of the listed organisms but repeating the data verbatim without checking its accuracy.

authority is unavoidable, because Darwinist educators cannot afford to reveal that their theory rests squarely on what the *Policy Statement* calls philosophical beliefs that are not subject to scientific test and refutation.

Darwinist scientists *believe* that the cosmos is a closed system of material causes and effects, and they *believe* that science must be able to provide a naturalistic explanation for the wonders of biology that appear to have been designed for a purpose. Without assuming these beliefs they could not deduce that common ancestors once existed for all the major groups of the biological world, or that random mutations and natural selection can substitute for an intelligent designer. Neither of these foundational beliefs is empirically testable and, according to the California *Policy Statement*, neither belongs in the science classroom.

The Darwinists may have made a serious strategic error in choosing to pursue a campaign of indoctrination in the public schools. Previously, the high school textbooks said relatively little about evolution except that most scientists believe in it, which is hard to dispute. Serious examination of the scientific evidence was postponed until college, and was provided mostly to biology majors and graduate students. Most persons outside the profession had little opportunity to learn how much philosophy was being taught in the name of science, and if they knew what was going on they had no opportunity to mount an effective challenge.

The Darwinists themselves have changed that comfortable situation by demanding that the public schools teach a great deal more "about evolution." What they mean is that the public schools should try much harder to persuade students to believe in Darwinism, not that they should present fairly the evidence that is causing Darwinists so much trouble. What goes on in the public schools is the public's business, however, and even creationists are entitled to point out errors and evasions in the textbooks and teaching materials. Invocations of authority may work for a while, but eventually determined protestors will persuade the public to grant them a fair hearing on the evidence. As many more people outside the Biblical fundamentalist camp learn how deeply committed Darwinists are to opposing theism of any sort, and how little support Darwinism finds in the scientific evidence, the Darwinists may wish that they had never left their sanctuary.

Chapter Twelve

Science and Pseudoscience

KARL POPPER PROVIDES the indispensable starting point for understanding the difference between science and pseudoscience. Popper spent his formative years in early twentieth century Vienna, where intellectual life was dominated by science-based ideologies like Marxism and the psychoanalytic schools of Freud and Adler. These were widely accepted as legitimate branches of natural science, and they attracted large followings among intellectuals because they appeared to have such immense explanatory power. Acceptance of either Marxism or psychoanalysis had, as Popper observed,

> the effect of an intellectual conversion or revelation, opening your eyes to a new truth hidden from those not yet initiated. Once your eyes were thus opened you saw confirming instances everywhere: the world was full of verifications of the theory. Whatever happened always confirmed it. Thus its truth appeared manifest; and unbelievers were clearly people who did not want to see the manifest truth; who refused to see it, either because it was against their class interest,

or because of their repressions which were still 'un-analyzed' and crying aloud for treatment. . . . A Marxist could not open a newspaper without finding on every page confirming evidence for his interpretation of history; not only in the news, but also in its presentation—which revealed the class bias of the paper—and especially of course in what the paper did not say. The Freudian analysts emphasized that their theories were constantly verified by their 'clinical observations.'

Popper saw that a theory that appears to explain everything actually explains nothing. If wages fell this was because the capitalists were exploiting the workers, as Marx predicted they would, and if wages rose this was because the capitalists were trying to save a rotten system with bribery, which was also what Marxism predicted. A psychoanalyst could explain why a man would commit murder— or, with equal facility, why the same man would sacrifice his own life to save another. According to Popper, however, a theory with genuine explanatory power makes *risky* predictions, which exclude most possible outcomes. Success in prediction is impressive only to the extent that failure was a real possibility.

Popper was impressed by the contrast between the methodology of Marx or Freud on the one hand, and Albert Einstein on the other. Einstein almost recklessly exposed his General Theory of Relativity to falsification by predicting the outcome of a daring experiment. If the outcome had been other than as predicted, the theory would have been discredited. The Freudians in contrast looked only for confirming examples, and made their theory so flexible that everything counted as confirmation. Marx did make specific predictions—concerning the inevitable crises of capitalism, for example—but when the predicted events failed to occur his followers responded by modifying the theory so that it still "explained" whatever had happened.

Popper set out to answer not only the specific question of how Einstein's scientific method differed from the pseudoscience of Marx and Freud, but also the more general question of what "science" is and how it differs from philosophy or religion. The accepted model, first described systematically by Francis Bacon, conceived of science as an exercise in *induction*. Scientists were believed to formulate theories in order to explain pre-existing experimental data, and to verify their theories by accumulating additional supporting evidence. But skeptical philosophers—especially David Hume—had questioned whether a series of factual observations could really establish the validity of a general law. One thing may follow another again and again in our inevitably limited experience, but there is always the possibility that further observations will reveal exceptions that disprove the rule. This was no mere theoretical possibility: scientists had been stunned to see the apparently invulnerable edifice of Newtonian physics crumble when modern techniques made it possible to make new kinds of observations.

The validity of induction as a basis for science was not only philosophically insecure, it was also inaccurate, because scientists do not work as the induction model prescribes. In scientific practice the theory normally precedes the experiment or fact-gathering process rather than the other way around. In Popper's words, "Observation is always selective. It needs a chosen object, a definite task, an interest, a point of view, a problem." Without a theory, scientists would not know how to design experiments, or where to look for important data.

Popper's inspired contribution was to discard the induction model and describe science as beginning with an imaginative or even mythological conjecture about the world. The conjecture may be wholly or partly false, but it provides a starting point for investigation when it is stated with sufficient clarity that it can be criticized. Progress is made not by searching the world for confirming examples, which can always be found, but by searching out the falsifying evidence that reveals the need for a new and better explanation.

Popper put the essential point in a marvelous aphorism: "The wrong view of science betrays itself in the craving to be right." In some cases this craving results from the pride of a discoverer, who defends a theory with every artifice at his disposal because his professional reputation is at stake. For Marxists and Freudians, the craving came from the sense of security they gained from having a theory that seemed to make sense out of the world. People base their careers and their personal lives on theories like that, and they feel personally threatened when the theory is under attack. Fear leads such people to embrace uncritically any device that preserves the theory from falsification.

Popper proposed the falsifiability criterion as a test for distinguishing science from other intellectual pursuits, among which he included pseudoscience and metaphysics. These terms have caused some confusion, because in ordinary language we identify "science" as the study of a particular kind of subject matter, such as physics or biology, as opposed to (say) history or literature. Popper's logic implies that a theory's scientific status depends less upon its subject matter than upon the attitude of its adherents towards criticism. A physicist or a biologist may be dogmatic or evasive, and therefore unscientific in method, while a historian or literary critic may state the implications of a thesis so plainly that refuting examples are invited. Scientific methodology exists wherever theories are subjected to rigorous empirical testing, and it is absent wherever the practice is to protect a theory rather than to test it.

"Metaphysics"—a catch-all term by which Popper designated all theories that are not empirically testable—is also a confusing category. Many readers assumed that Popper was implying that metaphysics is equivalent to nonsense. That was the view of a fashionable philosophical school called "logical positivism," with which Popper was sometimes incorrectly identified. The logical positivists tried to judge all thinking by scientific criteria, and to that end classified statements as meaningful only to the extent that they could be verified. An unverifiable statement, such as that "adultery is immoral" was either meaningless noise or merely an expression of personal taste.

Popper strongly opposed logical positivism, because he recognized that to discard all metaphysics as meaningless would make all knowledge impossible, including scientific knowledge. Universal statements, such as very general scientific laws, are not verifiable. (How could we verify that entropy always increases in the cosmos as a whole?) Moreover, Popper believed that it is out of metaphysics—that is, out of imaginative conjectures about the world—that science has emerged. For example, astronomy owes an enormous debt to astrology and mythology. The point of scientific investigation is not to reject metaphysical doctrines out of hand, but to attempt where possible to transform them into theories that can be empirically tested.

Popper insisted that metaphysical doctrines are frequently meaningful and important. Although they cannot be tested scientifically, they can nonetheless be criticized, and reasons can be given for preferring one metaphysical opinion to another. Popper even credited pseudoscientists like Freud and Adler with valuable insights that might one day play their part in a genuine science of psychology. His criticism was not that their theories were nonsense, but merely that they were deluded in thinking that they could verify those theories by clinical examinations that always allowed them to find what they expected to find.

Because of all these complications, the falsifiability criterion does not necessarily differentiate natural science from other valuable forms of intellectual activity. Popper's contribution was not to draw a boundary around science, but to make some frequently overlooked points about intellectual integrity that are equally important for scientists and non-scientists. He tells us not to be afraid to make mistakes, not to cover up the mistakes we have made, and not to take refuge in the false security that comes from having a worldview that explains things too easily.

How does Darwinism fare if we judge the practices of Darwinists by Popper's maxims? Darwin was relatively candid in acknowledging that the evidence was in important respects not easy to reconcile with his theory, but in the end he met every difficulty with a rhetorical solution. He described *The Origin of Species* as "one long argument," and the point of the argument was that the common ancestry thesis was so logically appealing that rigorous empirical testing was not required. He proposed no daring experimental tests, and thereby started his science on the wrong road. Darwin himself established the tradition of explaining away the fossil record, of citing selective breeding as verification without acknowledging its limitations, and of blurring the critical distinction between minor variations and major innovations.

The central Darwinist concept that later came to be called the "fact of evolution"—descent with modification—was thus from the start protected from empirical testing. Darwin did leave some important questions open, including the relative importance of natural selection as a mechanism of change. The resulting arguments about the process, which continue to this day, distracted attention from the fact that the all-important central concept had become a dogma.

The central concept is all-important because there is no real distinction between the "fact" of evolution and Darwin's theory.

When we posit that the discontinuous groups of the living world were united in the remote past in the bodies of common *ancestors*, we are implying a great deal about the process by which the ancestors took on new shapes and developed new organs. Ancestors give birth to descendants by the same reproductive process that we observe today, extended through millions of years. Like begets like, and so this process can only produce major transformations by accumulating the small differences that distinguish offspring from their parents. Some shaping force must also be involved to build complex organs in small steps, and that force can only be natural selection. There may be arguments about the details, but all the basic elements of Darwinism are implied in the concept of ancestral descent.

We can only speculate about the motives that led scientists to accept the concept of common ancestry so uncritically. The triumph of Darwinism clearly contributed to a rise in the prestige of professional scientists, and the idea of automatic progress so fit the spirit of the age that the theory even attracted a surprising amount of support from religious leaders. In any case, scientists did accept the theory before it was rigorously tested, and thereafter used all their authority to convince the public that naturalistic processes are sufficient to produce a human from a bacterium, and a bacterium from a mix of chemicals. Evolutionary science became the search for confirming evidence, and the explaining away of negative evidence.

The descent to pseudoscience was completed with the triumph of the neo-Darwinian synthesis, and achieved its apotheosis at the centennial celebration of the publication of *The Origin of Species* in 1959 in Chicago. By this time Darwinism was not just a theory of biology, but the most important element in a religion of scientific naturalism, with its own ethical agenda and plan for salvation through social and genetic engineering. Julian Huxley was the most honored speaker at Chicago, and his triumphalism was unrestrained.

> Future historians will perhaps take this Centennial Week as epitomizing an important critical period in the history of this earth of ours—the period when the process of evolution, in the person of inquiring man, began to be truly conscious of itself. . . . This is one of the first public occasions on which it has been frankly faced that all aspects of reality are subject to evolution, from atoms and stars to fish

and flowers, from fish and flowers to human societies and values—indeed, that all reality is a single process of evolution. . . .

In the evolutionary pattern of thought there is no longer either need or room for the supernatural. The earth was not created, it evolved. So did all the animals and plants that inhabit it, including our human selves, mind and soul as well as brain and body. So did religion. . . .

Finally, the evolutionary vision is enabling us to discern, however incompletely, the lineaments of the new religion that we can be sure will arise to serve the needs of the coming era.

These propositions go far beyond anything empirical science can demonstrate, of course, and to sustain this worldview Darwinists had to resort to all the tactics that Popper warned truth-seekers to avoid. Their most important device is the deceptive use of the vague term "evolution."

"Evolution" in Darwinist usage implies a completely naturalistic metaphysical system, in which matter evolved to its present state of organized complexity without any participation by a Creator. But "evolution" also refers to much more modest concepts, such as microevolution and biological relationship. The tendency of dark moths to preponderate in a population when the background trees are dark therefore demonstrates evolution—and also demonstrates, by semantic transformation, the naturalistic descent of human beings from bacteria.

If critics are sophisticated enough to see that population variations have nothing to do with major transformations, Darwinists can disavow the argument from microevolution and point to *relationship* as the "fact of evolution." Or they can turn to biogeography, and point out that species on offshore islands closely resemble those on the nearby mainland. Because "evolution" means so many different things, almost any example will do. The trick is always to prove one of the modest meanings of the term, and treat it as proof of the complete metaphysical system.

Manipulation of the terminology also allows natural selection to appear and disappear on command. When unfriendly critics are absent, Darwinists can just assume the creative power of natural selection and employ it to explain whatever change or lack of change has been observed. When critics appear and demand empirical confirmation, Darwinists can avoid the test by responding that scientists are discovering alternative mechanisms, particularly at the molecular level, which relegate selection to a less important role. The fact of evolution therefore remains unquestioned, even if there is a certain amount of healthy debate about the theory. Once the critics have been distracted, the Blind Watchmaker can reenter by the back door. Darwinists will explain that no biologist doubts the importance of Darwinian selection, because nothing else was available to shape the adaptive features of the phenotypes.

When disconfirming evidence cannot be ignored altogether, it is countered with ad hoc hypotheses. Douglas Futuyma's textbook tells college students that "Darwin more than anyone else extended to living things . . . the conclusion that mutability, not stasis, is the natural order." So he did, and in consequence paleontologists overlooked the prevalence in the fossil record of stasis. Stasis could not come to public notice until it was dressed up as evidence for "punctuated equilibrium," which sounded at first like a new theory but turned out to be a minor variant of Darwinism. Darwinists can also explain away stasis as an effect of stabilizing selection, or developmental constraints, or mosaic evolution—and so, like mutability, it is just what a Darwinist would expect.

Darwinists sometimes find confirming evidence, just as Marxists found capitalists exploiting workers and Freudians analyzed patients who said that they wanted to murder their fathers and marry their mothers. They find further instances of microevolution, or additional examples of natural relationships, or a fossil group that might have contained an ancestor of modern mammals. What they never find is evidence that contradicts the common ancestry thesis, because to Darwinists such evidence cannot exist. The "fact of evolution" is true by definition, and so negative information is uninteresting, and generally unpublishable.

If Darwinists wanted to adopt Popper's standards for scientific inquiry, they would have to define the common ancestry thesis as an empirical hypothesis rather than as a logical consequence of the fact of relationship. The pattern of biological relationships—including the universal genetic code—does imply an element of *commonality*, which means only that it is unlikely that life evolved by chance on many different occasions. Relationships may come from common ancestors, or from predecessors which were transformed by some means other than the accumulation of small differences, or from some process altogether beyond the ken of our science. Common ancestry is a hypothesis, not a fact, no matter how strongly it appeals to a materialist's common sense. As a hypothesis it deserves our most respectful attention, which, in Popper's terms, means that we should test it rigorously.

We would do that by predicting what we would expect to find if the common ancestry hypothesis is true. Until now, Darwinists have looked only for confirmation. The results demonstrate how right Popper was to warn that "Confirmations should count only if they are the result of *risky predictions*." If Darwin had made risky predictions about what the fossil record would show after a century of exploration, he would not have predicted that a single "ancestral group" like the therapsids and a mosaic like *Archaeopteryx* would be practically the only evidence for macroevolution. Because Darwinists look only for confirmation, however, these exceptions look to them like proof. Darwinists did not predict the extreme regularity of molecular relationships that they now call the molecular clock, but this phenomenon became "just what evolutionary theory would predict"—*after* the theory was substantially modified to accommodate the new evidence.

When analyzed by Popper's principles, the examples Darwinists cite as confirmation look more like falsification. There is no need to press for a verdict now, however. If Darwinists were to restate common ancestry as a scientific hypothesis, and encourage a search for falsifying evidence, additional evidence would be forthcoming. The final judgment on Darwinism can safely be left to the deliberative processes of the scientific community—once that community has demonstrated its willingness to investigate the subject without prejudice.

Prejudice is a major problem, however, because the leaders of science see themselves as locked in a desperate battle against religious fundamentalists, a label which they tend to apply broadly to anyone who believes in a Creator who plays an active role in worldly affairs. These fundamentalists are seen as a threat to liberal freedom, and especially as a threat to public support for scientific research. As the creation myth of scientific naturalism, Darwinism plays an indispensable ideological role in the war against fundamentalism. For that reason, the scientific organizations are devoted to protecting Darwinism rather than testing it, and the rules of scientific investigation have been shaped to help them succeed.

If the purpose of Darwinism is to persuade the public to believe that there is no purposeful intelligence that transcends the natural world, then this purpose implies two important limitations upon scientific inquiry. First, scientists may not consider all the possibilities, but must restrict themselves to those which are consistent with a strict philosophical naturalism. For example, they may not study genetic information on the assumption that it may be the product of intelligent communication. Second, scientists may not falsify an element of Darwinism, such as the creative power of natural selection, until and unless they can provide an acceptable substitute. This rule is necessary because advocates of naturalism must at all times have a complete theory at their disposal to prevent any rival philosophy from establishing a foothold.

Darwinists took the wrong view of science because they were infected with the craving to be right. Their scientific colleagues have allowed them to get away with pseudoscientific practices primarily because most scientists do not understand that there is a difference between the scientific method of inquiry, as articulated by Popper, and the philosophical program of scientific naturalism. One reason that they are not inclined to recognize the difference is that they fear the growth of religious fanaticism if the power of naturalistic philosophy is weakened. But whenever science is enlisted in some other cause—religious, political, or racialistic—the result is always that the scientists themselves become fanatics. Scientists see this clearly when they think about the mistakes of their predecessors, but they find it hard to believe that their colleagues could be making the same mistakes today.

Exposing Darwinism to possible falsification would not imply support for any other theory, certainly not any pseudoscientific theory based upon a religious dogma. Accepting Popper's challenge is simply to take the first step towards understanding: the recognition of ignorance. Falsification is not a defeat for science, but a liberation. It removes the dead weight of prejudice, and thereby frees us to look for the truth.

Research
Notes

These notes provide a guide to the sources actually used in the writing of this book, and attempt to answer questions that may occur to scientists and other readers who are acquainted with the professional literature. For a complete bibliography, I recommend Kevin Wirth's unpublished manuscript *The Creation-Evolution Bibliography, Including Major Works from 1830 to the Present, With Annotations* (1990). Copies of this remarkable research guide can be obtained by writing to Kevin Wirth, 7411 Park Wood Court #203, Falls Church, VA 22042 (cost: $20.00).

Chapter One **The Legal Setting**

The official legal citation for the Supreme Court decision in Aguillard v. Edwards is 482 U.S. 578 (1987). The Louisiana statute is reprinted in the appendix to the federal Court of Appeals opinion in the same case, 765 F.2d 1251, 1258-59 (5th Cir. 1985). That decision was by a 3-judge panel of the Court of Appeals; the full court refused to grant an "en banc" rehearing, but only by a vote of 8-7. This action is reported at 778 F.2d 225, along with the lively dissenting opinion by Judge Gee and the baffled response by Judge Jolly, the author of the panel decision.

In *Edwards* the Supreme Court applied what it calls its three-pronged *Lemon* test (first announced in the 1971 decision in *Lemon v. Kurtzman*, 403

U.S. 602). This test says that a challenged statute comports with the First Amendment's Establishment Clause only if (1) the legislature had a secular purpose; (2) the statute's principal effect is not to advance or inhibit religion; and (3) the statute does not excessively entangle government with religion. This test has been much criticized, and the essential criticisms are covered in Justice Scalia's dissenting opinion in *Edwards*.

I provided my own analysis of this area of the law in my article "Concepts and Compromise in First Amendment Religious Doctrine," in volume 72 of the *California Law Review*, p. 817 (1984). My view is that the *Lemon* test is a device for rationalizing a decision after it has been made on other grounds, because its criteria are vacuous and manipulable.

Besides *Edwards*, there are two other evolution cases worth noting. In Epperson v. Arkansas, 339 U.S. 99 (1968), the Supreme Court held unconstitutional a 40-year-old, unenforced state statute which made it an offense "to teach the theory or doctrine that mankind ascended or descended from a lower order of animals." An earlier version of the balanced treatment legislation was held unconstitutional by federal district Judge Overton in McLean v. Arkansas Board of Education, 529 F.Supp. 1255 (E.D.Ark. 1982). Unlike the Supreme Court, Judge Overton tried to define "science." I discuss his opinion in Chapter Nine.

The official position paper of the National Academy of Sciences was published in 1984, with beautiful illustrations, under the title "Science and Creationism: A View from the National Academy of Sciences." Excerpts from this paper were used in the Academy's *amicus curiae* brief in the Supreme Court case.

Stephen Jay Gould commented upon the Supreme Court decision in his article "Justice Scalia's Misunderstanding," 5 *Constitutional Commentary* 1 (1988). Gould criticizes Scalia for taking an incorrect view of the nature of science and for writing that, on the record before it, the Court should not say that "the scientific evidence for evolution is so conclusive that no one would be gullible enough to believe that there is any real scientific evidence to the contrary." Gould responds: "But this is exactly what I, and all scientists, do say." Gould appeared not to understand a legal point that all the Justices took for granted: the courts may not find against a party on a disputed issue of fact (e.g. whether scientific evidence against evolution exists) without giving the party an opportunity to present its evidence and expert witnesses in a trial. The trial court had held the Louisiana statute unconstitutional because of its presumed religious purpose, without allowing the state an opportunity to show what kind of evidence creation-scientists would present in classrooms if given the opportunity. The

Supreme Court therefore would have had no basis for a finding that the evidence would be bogus or nonexistent.

Colin Patterson's 1981 lecture was not published, but I have reviewed a transcript and Patterson restated his position, which I would label "evolutionary nihilism," in an interview with the journalist Tom Bethell. (See Bethell, "Deducing from Materialism," *National Review*, Aug. 29, 1986, p. 43.) I discussed evolution with Patterson for several hours in London in 1988. He did not retract any of the specific skeptical statements he has made, but he did say that he continues to accept "evolution" as the only conceivable explanation for certain features of the natural world.

Irving Kristol's essay "Room for Darwin and the Bible" appeared in *The New York Times* op-ed page for September 30, 1986. The title was unfortunate, because Kristol's thesis was not that the Bible should be included in science classes but that Darwinism should be taught less dogmatically. Stephen Jay Gould's reply essay appeared in the January 1987 issue of *Discover* magazine with the title "Darwinism Defined: The Difference between Fact and Theory."

The quotations attributed to Richard Dawkins are from his book *The Blind Watchmaker* (1986), and from his review in *The New York Times* of Donald Johanson and Maitland Edey's 1989 book *Blueprints*.

For accounts of the *Scopes* trial see Kevin Tierney's *Darrow: A Biography* (1979); L. Sprague de Camp's *The Great Monkey Trial* (1968); and Edward J. Larson's *Trial and Error: The American Controversy over Creation and Evolution* (rev. ed. 1989). The story is also nicely retold in Gould's essay "A Visit to Dayton," in *Hen's Teeth and Horse's Toes*, which relies upon Ray Ginger's 1958 book *Six Days or Forever*. This is as good a place as any to put on the record that I am an admirer of Gould's essays; despite a difference of outlook I nearly always profit from reading them. Perhaps he will feel that I did not profit enough. The story of Henry Fairfield Osborn and "Nebraska Man" is retold in Roger Lewin's *Bones of Contention* (1987).

The legal citation for the Tennessee Supreme Court's opinion is Scopes v. State, 154 Tenn. 105, 289 S.W. 363 (1927). In upholding the statute the court rejected an argument that prohibiting the teaching of evolution violated a clause of the state constitution which required the legislature "to cherish literature and science." The court reasoned that the legislature might have thought that "by reason of popular prejudice, the cause of education and the study of science generally will be promoted by forbidding the teaching of evolution in the schools of the state." One could thus argue that the statute in *Scopes* met the "secular purpose" requirement of

Edwards because the legislature had the secular purpose of obtaining public support for a science curriculum.

Chapter Two   **Natural Selection**

The primary source for the defense of neo-Darwinist natural selection used in this chapter is Douglas Futuyma's 1983 book, *Science on Trial: The Case for Evolution*. This is the book most frequently cited to me by Darwinists as having made the most powerful case for Darwinism and against creationism. Futuyma does a particularly thorough job of marshalling the evidence, and his viewpoint is orthodox neo-Darwinism. The quotes in this chapter are from Futuyma's Chapter Six.

Futuyma is not just a polemicist, but the author of one of the leading college textbooks on evolution and an internationally recognized authority. The cover of *Science on Trial* records glowing tributes from Ernst Mayr, Richard Leakey, David Pilbeam, Ashley Montagu, and Isaac Asimov. The praise from Mayr ("Professor Futuyma has provided a masterly summation of the evidence for evolution . . . ") is especially important. Mayr is the most prestigious living Darwinist authority, a man of prodigious knowledge whose opinions virtually define orthodoxy in this field.

The quotations from Pierre Grassé are from the 1977 English translation of his book *Evolution of Living Organisms*, pp. 124–25, 130. This book was originally published in France in 1973 with the title *L'Evolution du Vivant*. Grassé was an evolutionist, but an anti-Darwinist. As we shall see in the next chapter, this viewpoint propelled him towards the heresy of vitalism, which Darwinists regard as little better than creationism. Dobzhansky's book review begins with the following tribute:

> The book of Pierre P. Grassé is a frontal attack on all kinds of "Darwinism." Its purpose is "to destroy the myth of evolution, as a simple, understood, and explained phenomenon," and to show that evolution is a mystery about which little is, and perhaps can be, known. Now one can disagree with Grassé but not ignore him. He is the most distinguished of French zoologists, the editor of the 28 volumes of *Traite de Zoologie*, author of numerous original investigations, and ex-president of the Academie des Sciences. His knowledge of the living world is encyclopedic.

It seems therefore that it is possible for a person in complete command of the facts to come to the conclusion that Darwinism is a myth. The concluding paragraph of Dobzhansky's review indicates the philosophical basis for the dispute between Grassé and the neo-Darwinists:

The mutation-selection theory attempts, more or less successfully, to make the causes of evolution accessible to reason. The postulate that the evolution is "oriented" by some unknown force explains nothing. This is not to say that the synthetic . . . theory has explained everything. Far from this, this theory opens to view a great field which needs investigation. Nothing is easier than to point out that this or that problem is unsolved and puzzling. But to reject what is known, and to appeal to some wonderful future discovery which may explain it all, is contrary to sound scientific method. The sentence with which Grassé ends his book is disturbing: "It is possible that in this domain biology, impotent, yields the floor to metaphysics."

But why is it not possible that the development of life may have required some orienting force which our science does not understand? To reject that possibility because it is "disturbing" is to imply that it is better to stick to a theory which is against the weight of the evidence than to admit that the problem is unsolved.

My discussion of artificial selection deals with the laboratory fruitfly breeding experiments only briefly, and this will no doubt occasion Darwinist protests. An experimenter can greatly increase or decrease the number of bristles in a fruitfly (this is Futuyma's prime example), or greatly reduce wing size, etc., but the fruitflies are still fruitflies, usually maladapted ones. Some accounts credit the fruitfly experiments with producing new species, in the sense of populations which do not breed with each other; others dispute that the species border has in reality been crossed. Apparently the question turns on how narrowly or broadly one defines a species, especially with respect to populations that are inhibited from interbreeding but not totally incapable of it. I am not interested in pursuing the question, because what is at issue is the capacity to create new organs and organisms by this method, not the capacity to produce separated breeding populations. In any case, there is no reason to believe that the kind of selection used in the fruitfly experiments has anything to do with how fruitflies developed in the first place.

Horticulturists have developed plant hybrids which can breed with each other but not with either parent species. See Ridley, *The Problems of Evolution* (1985), pp. 4–5. On the other hand, the ability to alter plants by selection is also limited by the genetic endowment of the species and ceases once that capacity for variation is exhausted.

The quotations in the "tautology" section are from Norman Macbeth's *Darwin Retried* (1971), pp. 63-64; *A Pocket Popper* (1983), pp. 242; and C. H. Waddington, "Evolutionary Adaptation," in *Evolution after Darwin*, vol. 1, pp. 381–402 (Tax, ed., 1960). The "deductive argument" quotes are from

Colin Patterson's *Evolution* (1978), p. 147, and A. G. Cairns-Smith's *Seven Clues to the Origin of Life,* (1985), p. 2.

Gould commented on the tautology issue and the analogy between artificial and natural selection in his essay "Darwin's Untimely Burial," in the collection *Ever Since Darwin.* This essay responded to a magazine article critical of Darwinism by Tom Bethell, and both papers are re-printed in the reader *Philosophy of Biology* (Ruse, ed., 1989). Gould con-ceded that the tautology criticism "applies to much of the technical literature in evolutionary theory, especially to the abstract mathematical treatments that consider evolution only as an alteration in numbers, not a change in quality." He argued, however, that "superior design in changed environments" is a criterion of fitness independent of the fact of differen-tial survival, and therefore the theory as Darwin formulated it is not a tautology. I agree that in principle natural selection can be formulated non-tautologically, as in Kettlewell's industrial melanism experiment. The problem is not that the theory is inherently tautological, but rather that the absence of evidence for the important claims Darwinists make for natural selection continually tempts them to retreat to the tautology. In Chapter Four we will see that Gould himself explains the survival of species as due to their possessing the quality of "resistance to extinction."

In raising the tautology issue I am not merely taking advantage of a few careless statements. When the critics are not watching, Darwinists continue to employ natural selection in its tautological form as the self-evident explanation for whatever change or lack of change happened to occur. The important point is that the Darwinists have been tempted continually by the thought that their theory could be given the status of an *a priori* truth, or a logical inevitability, so that it could be known to be true without the need of empirical confirmation. Their susceptibility to this temptation is understandable. When the theory is stated as a hypothesis requiring empirical confirmation, the supporting evidence is not impressive.

For an excellent review of the tautology issue and the flaws in the arguments for natural selection as a creative force, see R. H. Brady's "Dogma and Doubt," in the *Biological Journal of the Linnaen Society* (1982); 17: 79-96.

Kettlewell's observation of industrial melanism in the peppered moth (Biston betularia) has been cited in countless textbooks and popular trea-tises as proof that natural selection has the kind of creative power needed to produce new kinds of complex organs and organisms. The 1990 *Science Framework* published by the California State Board of Education to guide textbook publishers (see Chapter Eleven for an analysis of its contents) has tried to correct the misrepresentation:

> Students should understand that this is not an example of evolutionary change from light-colored to dark-colored to light-colored moths, because both kinds were already in the population. This is an example of natural selection, but in two senses. First, temporary conditions in the environment encouraged selection against dark-colored moths and then against light-colored moths. But second, and just as important, is the selection to maintain a balance of both black and white forms, which are adaptable to a variety of environmental circumstances. This balanced selection increases the chances for survival of the species. This is in many ways the most interesting feature of the evolution of the peppered moth but one that is often misrepresented in textbooks. [p. 103.]

It is not difficult to understand why this frequent misrepresentation has occurred. Properly understood, industrial melanism illustrates natural selection as a fundamentally *conservative* force, which induces some relatively trivial variation within the species boundary but which also conserves the original genetic endowment so population frequencies can shift in the other direction when conditions change again. Such a process does not produce permanent, irreversible change of the kind required to produce new species, let alone new phyla. What the textbook writers have wanted to illustrate, however, is a process of natural selection capable of producing an insect from a microbe, a bird from a reptile, and a man from an ape. Suppressing the conservative implications of industrial melanism was necessary to achieve that objective.

How do Darwinists explain the apparent contradiction between natural selection and sexual selection? Mayr's essay "An Analysis of the Concept of Natural Selection," notes that sexual selection came back to prominence after the commemoration of the centennial of *The Descent of Man* in 1971. He concedes that "the existence of selfish selection for reproductive success poses a dilemma for the evolutionary biologist," because it tends to make the species less fit for survival and may even lead to extinction. Natural selection is not expected to achieve perfection, however, and the frequency of extinction itself shows that selection does not necessarily find an appropriate answer to every problem. See Mayr, *Toward a New Philosophy of Biology* (1988), pp. 105–06. Dawkins, who devotes several pages of *The Blind Watchmaker* to sexual selection asks "Why shouldn't fashion [in female sexual taste] coincide with utility?" He makes no attempt to answer, other than to show that, however the anti-utilitarian female preference arose, the force of sexual selection would tend to preserve it. (p. 205)

In his second classic, *The Descent of Man*, Darwin came close to repudiating the theory of natural selection as he had stated in *The Origin of Species*:

> A very large yet undefined extension may safely be given to the direct and indirect results of natural selection; but I now admit . . . that in the earlier editions of my "Origin of Species" I probably attributed too much to the action of natural selection or the survival of the fittest. . . . I had not formerly sufficiently considered the existence of many structures which appear to be, as far as we can judge, neither beneficial nor injurious; and this I believe to be one of the greatest oversights as yet detected in my work. I may be permitted to say as some excuse, that I had two distinct objects in view, firstly, to show that species had not been separately created, and secondly, that natural selection had been the chief agent of change, though largely aided by the inherited effects of habit, and slightly by the direct action of the surrounding conditions. Nevertheless, I was not able to annul the influence of my former belief, then widely prevalent, that each species had been purposely created; and this led to my tacitly assuming that every detail of structure, excepting rudiments, was of some special, though unrecognized, service. . . . If I have erred in giving to natural selection great power, which I am far from admitting, or in having exaggerated its power, which is in itself probable, I have at least, as I hope, done good service in aiding to overthrow the dogma of separate creations. [Darwin, *The Descent of Man*, quoted in Himmelfarb, *Darwin and the Darwinian Revolution* (1959), p. 302.]

Himmelfarb remarks upon "the alternating rhythm of self-recrimination and self-extenuation" in this curious statement. Darwin's explanation for having exaggerated the importance of natural selection is particularly intriguing, because he had no lingering attachment to creationism in 1859, and any overstatement would have been motivated by a desire to make the case *against* creation as powerful as possible. The passage almost implies that natural selection was a rhetorical device, important mainly for building the case against creationism, which could be re-evaluated and downgraded once its purpose had been served.

The quotation from Julian Huxley is from page 50 of *Evolution in Action* (1953).

Chapter Three   **Mutations Great and Small**

Darwin's letter to Charles Lyell is quoted on p. 249 of Dawkins' *The Blind Watchmaker*. Dawkins goes on to comment: "This is no petty matter. In Darwin's view, the whole point of the theory of evolution by natural selection was that it provided a non-miraculous account of the existence of complex adaptations."

Darwin's "uncompromising philosophical materialism" is the subject of the first two essays in Gould's collection *Ever Since Darwin*. Gould points out that "Other evolutionists spoke of vital forces, directed history, organic striving, and the essential irreducibility of mind—a panoply of concepts that traditional Christianity could accept in compromise, for they permitted a Christian God to work by evolution instead of creation. Darwin spoke only of random variation and natural selection." (pp. 24–25.) Gould also thinks that Darwin's turn to materialism may have been partly a reaction against the religious fundamentalism of the overbearing Captain Fitzroy, whose conversation he endured for five years on the *Beagle*. "Fitzroy may well have been far more important than finches, at least for inspiring the materialistic and antitheistic tone of Darwin's philosophy and evolutionary theory." (p. 33.)

Gould's candid portrayal of the role that philosophical preference and even personal prejudice may have played in Darwin's theorizing is refreshing, because the impression is often given that Darwin was a devout creationist who developed his theory only because of the irresistible pressure of the empirical evidence. Darwin's indifference to the empirical objections to gradualism offered by T. H. Huxley and others shows how false this picture is. Like his friend Charles Lyell, the founder of uniformitarian geology, Darwin was sure the evidence must be misleading when it led in a direction contrary to his philosophy. See also Gould's fascinating essay on Lyell, which observes that "To circumvent this literal appearance [of geologic catastrophes], Lyell imposed his imagination upon the evidence. The geologic record, he argued, is extremely imperfect and we must interpolate into it what we can reasonably infer but cannot see." (*Ever Since Darwin*, p. 150.) As we shall see in the next chapter, Darwin took this example much to heart.

Gertrude Himmelfarb's biography of Darwin is revealing on the question of his religious inclinations (and on other subjects as well). Darwin's father Robert was a secret unbeliever who maintained a facade of orthodoxy so thorough that it included planning a clerical career for Charles. According to Himmelfarb:

> Although Robert's mode of expressing, or rather suppressing, his disbelief did not commend itself to his son, the knowledge of that disbelief may have been of some influence. Not only did it make disbelief, when it came, appear to be a natural, acceptable mode of thought, so that loss of faith never presented itself to him as a moral crisis or rebellion; more than that, it seemed to enjoin disbelief precisely as a filial duty. One of the passages which was deleted from the autobiography explained why Charles not only

could not believe in Christianity but would not wish to believe in it. Citing the 'damnable doctrine' that would condemn all disbelievers to eternal punishment, he protested that 'this would include my Father, Brother, and almost all my best friends'—which made it an unthinkable, to say nothing of thoroughly immoral, idea. There may be more sophisticated reasons for disbelief, but there could hardly have been a more persuasive emotional one. (p. 22.)

This sort of information should not lead anyone into the "genetic fallacy," by which a theory is held to be wrong if caused by irrational factors. The correct conclusion to be drawn is merely that Darwinism should not be excused from the rigorous empirical testing which science requires of other theories.

For the orthodox Darwinist position on the evolution of complex organs this chapter relies on Ernst Mayr and Richard Dawkins. Dawkins' book *The Blind Watchmaker* is devoted primarily to this subject, and Dawkins is so brilliant an advocate that a reader can easily overlook (as most reviewers have) the absence of evidence for some of the critical points. For the quotations see pages 81, 84, 85–86, 89–90, 93, 230–33, 249. The Ernst Mayr quotations are from his 1988 collection *Toward a New Philosophy of Biology*: see pages 72, 464–66.

For Gould on Goldschmidt (some detractors refer to the pair as "Gouldschmidt") see "The Return of the Hopeful Monster" in the collection *The Panda's Thumb*. Gould's "new and general theory" paper has been reprinted in the collection *Evolution Now: A Century After Darwin* (Maynard Smith, ed., 1982). Those who want to read Goldschmidt in his own words are advised to look at his 1952 article in the journal *American Scientist* (vol. 40, p. 84), rather than his very detailed 1940 volume *The Material Basis of Evolution*, which is based on the Silliman Memorial Lectures he gave at Yale in 1939.

The Wistar Institute symposium is reported in *Mathematical Challenges to the Neo-Darwinian Interpretation of Evolution* (P. S. Moorehead and M. M. Kaplan, ed., 1967). The Darwin quotes are from the *The Origin of Species*, pp. 142, 219–20 (Penguin Library 1982).

The accepted theory of mutation is currently under challenge from an unexpected quarter. Researchers at the Harvard School of Public Health published a paper in *Nature* in 1988 (vol. 335, p. 142), reporting experimental evidence that some bacteria can produce directed helpful mutations in response to a change in their environment. If these preliminary indications were substantiated in a wider context an entirely new theory of mutation might arise in place of the neo-Darwinist theory that mutations are random and directionless. Conceivably this might lead to a new theory of evolution more in line with the views of Goldschmidt and Grassé than with neo-Darwinism, but for now no one knows how to account for a mystery like guided mutations and mainstream science will understandably require a great deal of evidence before accepting that such a phenomenon is of general significance.

Chapter Four **The Fossil Problem**

Gould's essay "The Stinkstones of Oeningen," in the collection *Hen's Teeth and Horse's Toes*, provides a good short introduction to the science of Georges Cuvier. Gould displays here the sympathetic understanding that often graces his historical sketches. Cuvier's reputation is in eclipse today, but in his time he was known as the Aristotle of biology, the virtual founder of the modern sciences of anatomy and paleontology, and a major statesman and public figure. Gould thoroughly refutes the prejudice that Cuvier's belief in catastrophes and the fixity of species was rooted in religious prejudice; on the contrary, Cuvier was far less committed to *a priori* philosophical principles than Lyell and Darwin.

Cuvier believed that evolution was impossible because an animal's major organs are so interdependent that a change in one part would require simultaneous changes in all the others—an impossible systemic macromutation. Gould comments parenthetically: "We would not deny Cuvier's inference today, but only his initial premise of tight and ubiquitous correlation. Evolution is mosaic in character, proceeding at different rates in different structures. An animal's parts are largely dissociable, thus permitting historical change to proceed." I suspect that this conclusion is based not on experimental proof, but upon wishful thinking—"this must be true or evolution couldn't have happened." Gould's remark does suggest a way in which the hypothesis of "mosaic evolution" could be tested, by transplanting organs from one kind of animal into another.

Darwin expected Charles Lyell to come around eventually and endorse his theory. After listing in the first edition of *The Origin of Species* all the distinguished paleontologists and geologists who "maintained the immutability of species," he added that "I have reason to believe that one great authority, Sir Charles Lyell, from further reflexion entertains grave doubts on this subject." Himmelfarb's biography reports that, when Lyell failed to give an unequivocal endorsement of evolution in a work published in 1863, "Darwin's disappointment amounted almost to a sense of betrayal." Lyell announced his conversion to mutability in a later edition of the same work in 1867, perhaps out of genuine conviction and perhaps out of a combination of friendship and an unwillingness to be left behind.

The Darwin quotations are from the first edition of *The Origin of Species* (Penguin Library edition, 1982), pages 133, 205, 292–93, 301–02, 305, 309, 313, 316, 322.

Louis Agassiz is the model of what happened to scientists who tried to resist the rising tide of evolution. Agassiz's tragedy is described in Gould's essay "Agassiz in the Galapagos," in *Hen's Teeth and Horse's Toes*. As Gould tells it, the Swiss-born Harvard professor was "without doubt, the greatest and most influential naturalist of nineteenth-century America," a great scientist and a social lion who was an intimate of just about everyone who mattered. "But Agassiz's summer of fame and fortune turned into a winter of doubt and befuddlement," because his idealist philosophical bias prevented him from embracing Darwin's theory. All his students became evolutionists and he had long been a sad and isolated figure when he died in 1873. I agree that Agassiz's philosophical bias was strong, but no stronger than the uniformitarian bias of Lyell and Darwin, and it may be that his incomparable knowledge of the fossil evidence was more important in restraining him from embracing a theory that relied so heavily upon explaining away that evidence. Ironically, Agassiz's best-remembered work, the *Essay on Classification*, was published in 1859, now remembered as the year of *The Origin of Species*.

Futuyma's dismissal of Agassiz illustrates how eagerly the Darwinists accepted a single fossil intermediate as proving their case: "The paleontologist Louis Agassiz insisted that organisms fall into discrete groups, based on uniquely different created plans, between which no intermediates could exist. Only a few years later, in 1868, the fossil *Archaeopteryx*, an exquisite intermediate between birds and reptiles, demolished Agassiz's argument, and he had no more to say on the unique character of birds." Futuyma, *Science on Trial*, p. 38. Specific cases of fossil intermediates are discussed in Chapter Six.

Douglas Dewar, a leader of the English Creation Protest Movement of the 1930s, described Darwinist bias in terms that foreshadow the punctuationalist critique of today. He wrote that biologists "allowed themselves to be dominated by the philosophical concept of evolution. They gave the hypothesis a warm welcome and set themselves to seek evidence in its favor. . . .[When some favorable evidence was found] it is not surprising that the hypothesis became generally accepted by biologists. It was perhaps but natural that they in their enthusiasm should regard the theory not merely as a most useful working hypothesis but as a law of nature. In the eighties of the last century we find the President of the American Association, Professor Marsh, saying: 'I need offer no argument for evolution, since to doubt evolution is to doubt science, and science is only another name for truth.' After the adoption of this attitude an evolutionary interpretation was put on every discovery. Facts that did not appear to fit in with the theory were regarded as puzzles that would eventually be solved." Dewar, *Difficulties of the Evolution Theory* (1931), pp. 2–3.

Gould's 1989 book *Wonderful Life* provides a splendid description of the Cambrian explosion and of the "Burgess Shoehorn," one of many efforts by paleontologists to provide a description of the fossil evidence consistent with their Darwinist preconceptions. Gould's remarks about the artifact theory and its demise are from pp. 271–73. Gould also reports on the current status of the dispute over the Ediacaran fauna at pp. 58–60 and 311–14. See also his essay "Death and Transfiguration," in the collection *Hen's Teeth and Horse's Toes*.

Gould's philosophical thesis in *Wonderful Life* is the least interesting thing about the book, although it has received a great deal of ·publicity. He speculates that evolution couldn't be expected to produce the same outcome (i.e. humans) a second time, because it proceeds by fortuitous factors rather than by deterministic laws. The picture of evolution as progress leading inevitably to "higher" forms of life like ourselves has been attractive to many Darwinists, and has helped to make evolution palatable to theists as a naturalistic version of a divine plan. It seems to me that a theist could take Gould's scientific description and draw the conclusion that a guiding creative intelligence outside nature had to be involved, because the creation of mankind (or insects, for that matter) is inexplicable without some powerful directional force to force life into patterns of greater complexity.

Steven M. Stanley's theory of evolution by rapid branching is presented for the general reader in his book *The New Evolutionary Timetable* (1981). The quotations in this chapter are from pages 71, 93–95, 104.

Eldredge and Gould's 1972 paper, "Punctuated Equilibria, an Alternative to Phyletic Gradualism," is reprinted as the appendix to Eldredge's book *Time Frames*. This book is the source of most of the Eldredge quotes in the chapter: pp. 59, 144–45. The longest quote is from his paper "Evolutionary Tempos and Modes: A Paleontological Perspective," in the collection *What Darwin Began: Modern Darwinian and Non-Darwinian Perspectives on Evolution* (Godfrey, ed., 1985). Chapter Three of *Time Frames* gives a good introductory description of the basic dilemma of paleontology, which is whether to read the fossil evidence in its own terms (example:

Schindewolf), or to stick to an interpretation acceptable to Darwinists (example: Simpson).

The basic description of punctuated equilibrium in the text is adapted from Gould's "The Episodic Nature of Evolutionary Change," in *The Panda's Thumb*. The very next essay in the collection is "The Return of the Hopeful Monster," which indicates why some people got the impression that punctuated equilibrium was a code term for "Goldschmidt-Schindewolf." The two T. H. Huxley theme quotes at the front of Gould and Eldredge's 1977 paper are: (1) to Darwin: "You have loaded yourself with an unnecessary difficulty in adopting *Natura non facit saltum* so unreservedly"; and (2) to the macromutationist William Bateson: "I see you are inclined to advocate the possibility of considerable 'saltus' on the part of Dame Nature in her variations. I always took the same view, much to Mr. Darwin's disgust."

That the charges of "Goldschmidtism" were not groundless can be readily documented from Gould's 1980 and 1984 papers. The 1980 "New and General Theory" paper argued the following thesis: (1) Richard Goldschmidt was right to conclude that speciation is a fundamentally different process from microevolution, requiring another kind of mutations. Gould termed this species barrier the "Goldschmidt break." (2) Speciation is random in direction compared to macroevolutionary trends, so that macroevolutionary trends are the result of differential success among species (*i.e.* "species selection," instead of natural selection among individual organisms as Darwin thought). "With apologies for the pun, the hierarchical rupture between speciation and macroevolutionary trends might be called the Wright break" [after Sewall Wright].[1] (3) The reproductive success of a species is not necessarily the result of adaptive advantages, but may be due to the fortuitous presence of an ecological niche, or to such factors as "high rates of speciation and strong resistance to extinction." With respect to the evolution of complex organs, Gould disavowed reliance on "saltational origin of entire new designs," but proposed instead "a potential saltational origin for the essential features of key adaptations."

For a neo-Darwinist response to Gould's paper see Stebbins and Ayala, "Is a new Evolutionary Synthesis Necessary?" in *Science*, vol. 213, p. 967 (August 1981). Their basic line is that the synthesis can incorporate any special features of macroevolution that "are compatible with the theories and laws of population biology." This qualification is extremely important, because the need for a separate theory of macroevolution arises from the

---

[1] Having committed himself to a pun, I do not know how Gould could have resisted adding that the species which thrive are the one that have the "Wright stuff."

fact that the theories of population biology are inadequate to account for macroevolution, if the fossil record problem is honestly faced rather than conjured away with ad hoc hypotheses.

Gould's explanation that the purpose of the punctuated equilibrium hypothesis was to permit the reporting of stasis is quoted from his essay "Cardboard Darwinism," in *The Urchin in the Storm*.

Ernst Mayr's opinion of the punctuated equilibrium controversy may be found in his 1988 essay, "Speciational Evolution through Punctuated Equilibria," in the collection of his papers titled *Toward a New Philosophy of Biology*. Mayr generally tries to put the most reasonable interpretation (from a neo-Darwinist perspective) on what Gould and Eldredge wrote. His most severe judgment is that "Nothing incensed some evolutionists more than the claims made by Gould and associates that they had been the first to have discovered, or at least to have for the first time properly emphasized, various evolutionary phenomena already widely accepted in the evolutionary literature." (p. 463.) For a livelier presentation of the same point of view, see the description of the controversy in Dawkins' *The Blind Watchmaker*.

Much of the controversy in paleontological circles over mass extinctions has been over whether the evidence supports theories such as that of Louis and Walter Alvarez. The Alvarez theory is that an asteroid struck the earth at the end of the Cretaceous era (the K-T boundary), causing a worldwide dust cloud which temporarily suppressed photosynthesis and thus disrupted the food chain. According to a 1982 review of the subject by Archibald and Clemens [*American Scientist*, vol. 70, p. 377], the paleontological evidence on the whole supports a more gradual pattern of extinction occurring over thousands or even millions of years. A 1988 article in *Science* (vol. 239, p. 729), reporting discussions at the annual meeting of the Geological Society of America, concluded that the pattern of extinctions occurred over thousands of years at the end of the Cretaceous period, but that the evidence for the asteroid theory is substantial and "the great impact at the boundary could indeed have sent a destabilized ecological system over the brink."

The question of whether the great extinctions were preceded by periods of more gradual extinction is the subject of ongoing research. According to a report in *Science* (11 January 1991, p. 160), new studies are showing that the dinosaurs and ammonites (ancient mollusks) were thriving up to the time of the asteroid impact. It is worth remarking that the only hard evidence Darwin cited in his passage arguing for gradual extinctions was the "wonderfully sudden" extermination of the ammonites.

A good brief account of the current state of research by science writer Richard Kerr appeared in *The Los Angeles Times* for June 12, 1989, part II, p. 3 (reprinted from *The Washington Post*). It seems safe to say that the predominant scientific opinion today is that a mass extinction at the K-T boundary occurred, caused by an asteroid or comet impact. A minority of geologists credit the mass extinction to volcanic activity, and many paleontologists continue to insist on a gradualist explanation for extinctions. Of course, it is difficult to determine when extinctions occurred with any precision, especially if the fossil record is anywhere near as imperfect as it has to be for Darwinism to be a serious possibility. Even if the mass extinctions occurred over many years as a result of climate changes, receding oceans, or whatever, the pattern would not necessarily be consistent with the gradual obsolescence postulated by Darwin.

On the issue of whether science textbooks and other sources have been presenting a distorted picture of the fossil record both to the general public and to the scientific profession, a letter published in *Science* in 1981 by David Raup is of additional interest. Raup, based at the University of Chicago and the Field Museum, is one of the world's most respected paleontologists. The letter contains the passage:

> A large number of well-trained scientists outside of evolutionary biology and paleontology have unfortunately gotten the idea that the fossil record is far more Darwinian than it is. This probably comes from the oversimplification inevitable in secondary sources: low-level textbooks, semi-popular articles, and so on. Also, there is probably some wishful thinking involved. In the years after Darwin, his advocates hoped to find predictable progressions. In general, these have not been found—yet the optimism has died hard, and some pure fantasy has crept into textbooks. . . . One of the ironies of the evolution-creation debate is that the creationists have accepted the mistaken notion that the fossil record shows a detailed and orderly progression and they have gone to great lengths to accommodate this 'fact' in their Flood geology. [*Science*, vol. 213, p. 289.]

Raup's letter also comments that "Darwinian theory is just one of several biological mechanisms proposed to explain the evolution we observe to have happened." The question, however, is whether any mechanism other than Darwinian selection has been proposed which can both account for the development of complex systems and also satisfy the requirements of the population geneticists.

Raup's essay on the fossil record issue in Godfrey's *Scientists Confront Creationism* collection is particularly interesting. In what was supposed to be a polemic against creationism he included the following paragraph:

Darwin predicted that the fossil record should show a reasonably smooth continuum of ancestor-descendant pairs with a satisfactory number of inter-mediates between major groups. Darwin even went so far as to say that if this were not found in the fossil record, his general theory of evolution would be in serious jeopardy. Such smooth transitions were not found in Darwin's time, and he explained this in part on the basis of an incomplete geologic record and in part on the lack of study of that record. We are now more than a hundred years after Darwin and the situation is little changed. Since Darwin a tremendous expansion of paleontological knowledge has taken place, and we know much more about the fossil record than was known in his time, but the basic situation is not much different. *We actually may have fewer examples of smooth transitions than we had in Darwin's time, because some of the old examples have turned out to be invalid when studied in more detail.* To be sure, some new intermediate or transitional forms have been found, particularly among land vertebrates. But if Darwin were writing today, he would still have to cite a disturbing lack of missing links or transitional forms between the major groups of organisms. [Emphasis added.]

Raup went on to explain that evolutionists explain the disturbing lack of evidence in three ways: (1) Because of the nature of the classification system creatures have to be put in one group or another, and so the absence of intermediates is to some extent an artifact of taxonomic prac-tice; (2) The fossil record is still incomplete; and (3) Evolution may occur rapidly by punctuated equilibrium. Raup's conclusion: "With these con-siderations in mind, one must argue that the fossil record is compatible with the predictions of evolutionary theory." (From Godfrey, ed., pp. 156–58.) I think that the phrasing of that conclusion hints at a certain lack of conviction.

For a scholarly comparison of the evolutionary theories of Schindewolf and Simpson, see Marjorie Grene's article "Two Evolutionary Theories," in *The British Journal for the Philosophy of Science*, vol. 9, pp. 110–27, 185–93. Grene concludes that Schindewolf's theory was the more adequate of the two because Simpson's Darwinist reductionism caused him to "overlook essential aspects of the phenomena," and in general to try to avoid employ-ing embarrassing concepts that were nonetheless unavoidable and there-fore tended to creep back into his analysis in concealed form. Raup has described Schindewolf, who died in 1972, as "the most respected scholar of the fossil record in Germany and perhaps the world, widely known for his research on the great mass extinction at the end of the Permian period, 250 million years ago." Schindewolf was the first expert to suggest an extraterrestrial cause for mass extinctions. (Raup, *The Nemesis Affair*, p. 38.)

Chapter Five   **The Fact of Evolution**

Darwin's argument from classification is from Chapter 13 of *The Origin of Species*. The term "homology" was first used by Darwin's rival Richard Owen, the founding director of the British Natural History Museum. It is derived from the Greek word for agreement. Darwin defined "homology" in the 6th edition of *The Origin of Species* as "that relation between parts that results from their development from corresponding embryonic parts." According to a 1971 monograph by Sir Gavin De Beer, former Director of the British Natural History Museum and a renowned authority on embryology, "This is just what homology is not."

De Beer reported that "correspondence between homologous structures cannot be pressed back to similarity of positions of the cells of the embryo or the parts of the egg out of which these structures are ultimately differentiated." Moreover, "homologous structures need not be controlled by identical genes, and homology of phenotypes does not employ similarity of genotypes." De Beer rhetorically demanded to know:

> What mechanism can it be that results in the production of homologous organs, the same "patterns," in spite of their *not* being controlled by the same genes? I asked this question in 1938, and it has not been answered.

It is amusing to see De Beer, one of the most dogmatic of all the neo-Darwinists, sounding on this occasion like another Richard Goldschmidt. If homology actually reflects biological descent it ought to involve common embryonic parts and homologous genes, which is precisely why Darwin defined the term as he did. De Beer's monograph *Homology: An Unsolved Problem* is rarely mentioned, probably because unsolvable problems are not "interesting." Its main points are summarized in the chapter on homology in Denton's *Evolution: A Theory in Crisis*.

The remark "Nothing in biology makes sense except in the light of evolution" is the title of a famous lecture by Theodosius Dobzhansky. It is quoted in virtually every Darwinist apologetic as a decisive argument in favor of the theory.

The Gould quotes are from the essay "Evolution as Fact and Theory," in the collection *Hen's Teeth and Horse's Toes*. Gould makes substantially the same arguments in his reply to Irving Kristol, which is described in other respects in Chapter One. I use Gould as a starting point because he makes the case succinctly and as persuasively as anyone. Futuyma as usual does the best job of collecting the evidence; his important points are covered in this chapter or in other chapters. The Futuyma quote in this chapter is from page 48 of *Science on Trial*.

The Mark Ridley quote about how universal evolution is proved by microevolution plus uniformitarianism is from his *Evolution and Classification*. Ridley makes the same argument in the first chapter of *Problems of Evolution*. The quotation about the vertebrate sequence by Louis Agassiz is from the concluding pages of his *Principles of Zoology*, published in 1866.

Chapter Six **The Vertebrate Sequence**

The primary source for the information about the vertebrate fossil record in this chapter is Barbara J. Stahl's comprehensive text *Vertebrate History: Problems in Evolution* (Dover 1985), especially Chapters Five and Nine.

The information about the coelacanth and the rhipidistians is from Stahl, pp. 121–48; see also Denton, pp. 179–80, and a fine article by Max Hall (in the January 1989 *Harvard Magazine*) titled "The Survivor," with beautiful illustrations. The coelacanths and rhipidistians are classified together as crossopterygian fishes, and this last more general term is used in many texts and articles to describe the supposed ancestral group for amphibians. Stahl notes that the seymouriamorphs come too late in the fossil record to be reptile ancestors and in any event are now considered true amphibians, on pp. 238–39.

The comment by Gareth Nelson about how ancestors are picked is from an interview with journalist Tom Bethell published in *The Wall Street Journal* (December 9, 1986).

The discussion of the mammal-like reptiles is based upon Stahl (Chapter Nine), as well as the pertinent chapters in Futuyma and Grassé. The quote from Futuyma on this subject is from p. 85 of *Science on Trial* and the quote by Gould is from the "Evolution as Fact and Theory" essay discussed in Chapter Five. Following the example of other writers I have lumped the mammal-like reptiles together as "therapsids," avoiding the use of more specific technical terms— cynodonts, theriodonts, etc.—that would distract the general reader unnecessarily. The mammal-like reptiles are also sometimes called the *synapsida*, the subclass to which the group belongs. The essential point is that wherever one draws the line around the group of eligible ancestors for mammals, it contains a range of groups and numerous species, no particular one of which can be identified conclusively as ancestral to mammals. A quote from Grassé (p. 35) is helpful:

> All paleontologists note . . . that the acquisition of mammalian characteristics has not been the privilege of one particular order, but of *all the orders of theriodonts*, although to varying degrees. This progressive evolution toward

mammals has been most clearly noted in three groups of carnivorous theraps-
sids: the Therocephalia, Bauriamorpha and Cynodontia, each of which at
one time or another has been considered ancestral to some or all mammals.

James A. Hopson of the University of Chicago is a leading expert on the
mammal-like reptiles, and he argues the case for their status as mammal
ancestors in his article "The Mammal-like Reptiles: A Study of Transitional Fossils," in *The American Biology Teacher*, vol. 49, no. 1, p. 16 (1987).
Hopson is not testing the ancestry hypothesis in the sense that I do so in
this chapter, but attempting to show the superiority of the "evolution
model" to the creation-science model of Duane Gish. To that end he
demonstrates that therapsids can be arranged in a progressive sequence
leading from reptilian to mammalian forms, with the increasingly
mammal-like forms appearing later in the geological record. So far so
good, but Hopson does *not* present a genuine ancestral line. Instead he
mixes examples from different orders and subgroups, and ends the line in
a mammal (*Morganucodon*) which is substantially older than the therapsid
that precedes it. The proof may be good enough to make Hopson's specific
point, which is that for this example some form of evolutionary model is
preferable to the creation-science model of Gish, but his argument does
not qualify, or purport to qualify, as a genuine testing of the common
ancestry hypothesis in itself.

Futuyma defends *Archaeopteryx* as a transitional intermediate on pp.
188–89 of *Science on Trial*. Stahl notes in her text that "Since *Archaeopteryx*
occupies an isolated position in the fossil record, it is impossible to tell
whether the animal gave rise to more advanced fliers or represented only a
side branch from the main line." In the preface to the 1985 Dover edition,
she added the remark that "retrieval of true bird fossils of Lower Cretaceous age has only strengthened the argument that the famous feathered
Archaeopteryx may be an archaic side branch of the ancestral avian stock."
[pp. viii, 369.] Peter Wellnhofer's informative review article "*Archaeopteryx*"
appeared in the May 1990 issue of *Scientific American*. It does not take
account of Paul Sereno's announcement of the Chinese fossil bird discovery, which is reported in *The New York Times* for October 12, 1990.

Roger Lewin is a fine science writer who has written several books on
human evolution. For this chapter I relied particularly on his *Bones of
Contention* (1987). The two most prominent fossil discoverers, Donald Johanson and Richard Leakey, have also authored or co-authored informative books. For a brief overview of the whole subject, I recommend the
article by Cartmill, Pilbeam, and Isaac, "One Hundred Years of Paleoanthropology," in the *American Scientist*, vol. 74, p. 410 (1986).

There are two debunking accounts of the human evolution story from authors outside of mainstream science that deserve careful scrutiny. One is the privately printed *Ape-Men, Fact or Fallacy*, by Malcolm Bowden. Bowden is a creation-scientist, but unprejudiced readers will find his book thoroughly documented and full of interesting details. Bowden has an intriguing account of the Piltdown hoax, and like Stephen Jay Gould he concludes that the Jesuit philosopher and paleontologist Teilhard de Chardin was probably culpably involved in the fraud. Bowden persuaded me that there are grounds to be suspicious of both the Java Man and Pekin Man fossil finds, which established what is now called *Homo erectus*. The book is available from Sovereign Publications, P.O. Box 88, Bromley, Kent BR2 9PF, England. I would like to see the details he reports examined critically but fairly by unbiased scholars, but this is a pipedream.

The other non-mainstream debunking account is *The Bone Peddlers: Selling Evolution*, by William R. Fix. This book is marred for me by its later chapters, which accept evidence of parapsychological phenomena uncritically, but the chapters about the human evolution evidence are devastating. Fix opens with an account of a 1981 CBS television news story about presidential candidate Ronald Reagan's statement that the theory of evolution "is not believed in the scientific community to be as infallible as it once was believed." A spokesman for the American Association for the Advancement of Science responded that the 100 million fossils that have been identified and dated "constitute 100 million facts that prove evolution beyond any doubt whatever."

Stephen Stanley's *The New Evolutionary Timetable* provides an analysis of the hominid evidence in Chapter Seven. Stanley points out that the accepted hominid sequence is radically inconsistent with Dobzhansky's neo-Darwinist theory (in *Mankind Evolving*) that Australopithecine-to-man evolution occurred in a continuous lineage within a single gene pool. On the contrary, Stanley reports, there were a very small number of discrete, long-lived intermediate species that may have overlapped each other. Stanley proposes a model based on "rapidly divergent speciation."

The statements by Solly Zuckerman (now Lord Zuckerman) are from his 1970 book *Beyond the Ivory Tower*. Zuckerman returned to this subject in his 1988 autobiographical work *Monkeys, Men and Missiles*, where he recounted his "running debate" with Sir Wilfred Le Gros Clark on the interpretation of the australopithecines. Zuckerman believes that Le Gros Clark was "obsessed" with the subject and incapable of rational consideration of the evidence. No doubt the opinion was reciprocated.

Donald Johanson and Maitland Edey's popular book on the discovery of

A. *Afarensis, Lucy: The Beginnings of Mankind* (1981), does a good job of describing the main point at issue between Zuckerman and the anthropologists:

> To give Zuckerman his due, there were resemblances between ape skulls and australopithecine skulls. The brains were approximately the same size, both had prognathous (long, jutting) jaws, and so on. What Zuckerman missed was the importance of some traits that australopithecines had in common with men. Charles A. Reed of the University of Illinois had summarized Zuckerman's misunderstandings neatly in a review of the australopithecine controversy: "No matter that Zuckerman wrote of such characters as being 'often inconspicuous'; the important point was the presence of several such incipient characters in functional combinations. This latter point of view was one which, in my opinion, Zuckerman and his co-workers failed to grasp, even while they stated they did. Their approach was extremely static in that they essentially demanded that a fossil, to be considered by them to show any evidence of evolving toward living humans, must have essentially arrived at the latter status before they would regard it as having begun the evolutionary journey." In other words: if it wasn't already substantially human, it could not be considered to be on the way to becoming human. (p. 80)

This argument revealingly supports one of Zuckerman's main points, which was that attempts to place the fossils in an evolutionary sequence "depend . . . partly on guesswork, and partly on some preconceived conception of the course of hominid evolution." The *Australopithecines* possessed incipient characters, more visible to some eyes than to others, which might have developed into human features and which also might not have done so. If the fossil creatures were "on the way to becoming human," then the same was undoubtedly true of the disputed "incipient characters," but if they weren't then the characters were probably insignificant. The description of what the fossils were is influenced decisively by the preconception about what they were going to become.

Zuckerman's article "A Phony Ancestor," in *The New York Review of Books* for November 8, 1990, provides some additional comments in the course of a review of a book on the Piltdown fraud. He refers readers to an article he published in 1933 denying the "uniqueness of Peking Man" and suggesting that the hominids should be divided into two families containing: (1) Peking Man and Neanderthals; and (2) those with skulls like modern men. Zuckerman attributed the success of the Piltdown forgery to the fact that anthropologists deluded themselves in thinking that they could "diagnose with the unaided eye what they imagined were hominid characters in bones and teeth." He concluded that "The trouble is that they still do. Once committed to what their or someone else's eyes have told them, everything else has to accord with the diagnosis."

Zuckerman's biometric debunking of the *Australopithecines* occurred before the discovery of "Lucy" by Johanson. Lucy is a more primitive specimen of the genus than Dart's *A. Africanus*, and hence would be disqualified *a fortiori* if Zuckerman's conclusions about *Africanus* are correct. Although Johanson and his colleague Owen Lovejoy confidently assert that Lucy walked upright like a human, this claim has not gone unchallenged. The controversy is briefly summarized in Roger Lewin's *Human Evolution: An Illustrated Introduction*:

> Although Lucy's pelvis is most definitely not that of an ape, neither is it fully human in form, particularly in the angle of the iliac blades. Nevertheless, concludes Owen Lovejoy of Kent State University, biomechanical and anatomical studies of the mosaic pelvis indicate that the structure is consistent with a style of bipedality that is strikingly modern. By contrast, two researchers at the State University of New York at Stony Brook interpret the mixture of characters in Lucy's pelvis as indicative of a somewhat simian form of bipedality, a bent-hip, bent-knee gait. The difference of opinion is yet to be resolved.
>
> Studies on the Lucy skeleton and on other Hadar specimens show A. afarensis to have had long forelimbs and relatively short hindlimbs—an ape-like configuration. (Milford Wolpoff, of the University of Michigan, argues, however, that Lucy's small legs are the length one would expect in a human of her diminutive stature.) Even more ape-like are the distinctly curved finger and toe bones. The Stony Brook researchers, Randall Susman and Jack Stern, interpret these features as adaptations to significant arboreality. Others, including Lovejoy and White, suggest other interpretations might be possible. (p. 41.)

No doubt many interpretations are *possible*, but the hypothesis being tested in this chapter is that Lucy and the other hominids have been conclusively identified as human ancestors, without assistance from any presumption that the questioned ape-to-man transition must have occurred.

The "mitochondrial Eve" hypothesis and the resulting conflict between the molecular biologists and the physical anthropologists is given a good popular treatment (if one can overlook the vulgar writing style) in Michael H. Brown's *The Search for Eve* (Harper & Row, 1990). Brown seems unsure about whether his subject is science or imaginative fiction, and I think many readers will feel that his uncertainty is justified. The book shows the contempt that "hard science" molecular biologists have for the "softer" paleontologists who base their theories about human evolution upon reconstructions from isolated teeth, shattered skullcaps, and fragmented jaws. According to Allan Wilson's colleague Rebecca Cann: "Many paleontologists fear that if they expose the legitimate scientific limits of the certainty of their theories, fundamentalists and creation 'scientists' may misrepresent these data to dispute the fact that evolution occurred." (p. 239.)

Brown also quotes an interesting remark by Alan Mann, a professor of Paleoanthropology at the University of Pennsylvania: "Human evolution is a big deal these days. Leakey's world known, Johanson is like a movie star, women moon at him and ask for his autograph. Lecture circuit. National Science Foundation: big bucks. Everything is debatable, especially where money is involved. Sometimes people deliberately manipulate data to suit what they're saying." (p. 241.)

The *Basilosaurus* reconstruction is described for scientists in the article "Hind Limbs of Eocene *Basilosaurus:* Evidence of Feet in Whales," by Philip D. Gingerich, B. Holly Smith, and Elwyn L. Simons, in *Science*, vol. 249, pp. 154-57 (July 15, 1990). The article states that "Limb and foot bones described here were all found in direct association with articulated skeletons of *Basilosaurus isis* and undoubtedly represent this species." Although I accept the authors' description for purposes of this chapter, I confess that expressions like "found in direct association with" and "undoubtedly" whet my curiosity. Is it certain that *Basilosaurus* had shrunken hind limbs, or is it only certain that fossil foot bones were found reasonably close to *Basilosaurus* skeletons? The newspaper stories quote discoverer Philip Gingerich as saying that "I feel confident we can go back to any skeleton, measure out the distance from the head—about 40 feet—sweep away the sand, and find more feet." That is an admirably risky prediction, and if Gingerich can make it good, all doubts about who owned the feet should be put to rest.

Douglas Dewar, a creationist biologist who prominently dissented from the evolutionary orthodoxy in Britain in the 1930s, provided an amusing description of the problems involved in a hypothetical whale evolution scenario:

> Let us notice what would be involved in the conversion of a land quadruped into, first a seal-like creature and then into a whale. The land animal would, while on land, have to cease using its hind legs for locomotion and to keep them permanently stretched out backwards on either side of the tail and to drag itself about by using its forelegs. During its excursions in the water, it must have retained the hind legs in their rigid position and swim by moving

them and the tail from side to side. As a result of this act of self-denial we must assume that the hind legs eventually became pinned to the tail by the growth of membrane. Thus the hind part of the body would have become like that of a seal. Having reached this stage, the creature, in anticipation of a time when it will give birth to its young under water, gradually develops apparatus by means of which the milk is forced into the mouth of the young one, and meanwhile a cap has to be formed round the nipple into which the snout of the young one fits tightly, the epiglottis and laryngeal cartilage become prolonged downwards so as tightly to embrace this tube, in order that the adult will be able to breath while taking water into the mouth and the young while taking in milk. These changes must be effected completely before the calf can be born under water. Be it noted that there is no stage intermediate between being born and suckled under water and being born and suckled in the air. At the same time various other anatomical changes have to take place, the most important of which is the complete transformation of the tail region. The hind part of the body must have begun to twist on the fore part, and this twisting must have continued until the sideways movement of the tail developed into an up-and-down movement. While this twisting went on the hind limbs and pelvis must have diminished in size, until the former ceased to exist as external limbs in all, and completely disappeared in most, whales." [Quoted in Denton, pp. 217–18.]

Darwinists have concentrated almost entirely on animal evolution and have paid much less attention to the problems of macroevolution in plants, probably because this subject is not as relevant to the ascent of man. The 1971 monograph "The Mysterious Origin of Flower Plants," by Kenneth Sporne (Cambridge University Lecturer in Botany) comments:

Theories without number have been put forward concerning the origin and subsequent evolution of flowering plants, but none has received universal approval. Darwin, in a letter to Hooker, written in 1879, made the following comment: "The rapid development, as far as we can judge, of all the higher plants within recent geological times is an abominable mystery," and the situation has scarcely changed since then, in spite of the remarkable advances that have been made in the twentieth century.

Laurie Godfrey writes that paleobotanists have recently identified fossil pollens and leaves as "members of a primary adaptive radiation of angiosperms," in *Scientists Confront Creationism*, p. 201. I wish that paleobotanists would do for the plant evidence what I have tried to do for vertebrates, and test the common ancestry hypothesis by the plant fossil record. I suspect that the results would be embarrassing to Darwinists. Creationist sources frequently quote the remark of Cambridge University botanist E. Corner on the subject:

Much evidence can be adduced in favor of the theory of evolution—from biology, bio-geography and paleontology, but I still think that, to the unprejudiced, the fossil record of plants is in favor of special creation. If, however, another explanation could be found for this hierarchy of classification, it would be the knell of the theory of evolution. Can you imagine how an orchid, a duckweed, and a palm have come from the same ancestry, and have we any evidence for this assumption? The evolutionist must be prepared with an answer, but I think that most would break down before an inquisition. [From Corner's essay "Evolution," in *Contemporary Biological Thought*, see pages 95 and 97, (McLeod & Colby, ed., 1961).]

Chapter Seven   **The Molecular Evidence**

For background information on the molecular evidence I have relied principally upon three articles in the *Scientific American* magazine by prominent authorities: Motoo Kimura, "The Neutral Theory of Molecular Evolution" (Nov. 1979); G. Ledyard Stebbins and Francisco Ayala, "The Evolution of Darwinism" (July 1985); and Allan Wilson, "The Molecular Basis of Evolution" (October 1985).

The data regarding cytochrome c molecular sequence divergencies is from a table in Dayhoff's *Atlas of Protein Sequence and Structure*; it is reproduced in Denton's *Evolution: A Theory in Crisis* (1985). Denton pursues the thesis that the molecular evidence shows a world of discontinuous natural groupings that supports the essentialist or typological view rather than the Darwinist view of continuity over time. The Darwinist answer is to assume that the discontinuous groups of the present came about by continuous evolution from distant common ancestors. The question is whether the Darwinist assumption is merely a philosophical preference, or whether it is backed up by substantial evidence.

The quotation endorsing pan-selectionism by Ernst Mayr and the quotations attributed to Kimura are taken from Kimura's *Scientific American* article. Kimura acknowledges that to test the neutral theory "it is necessary to estimate such quantities as mutation rates, selection coefficients, population sizes, and migration rates" [over long stretches of geological time]. He concedes that "Many evolutionary biologists maintain that such population-genetic quantities can never be accurately determined and that consequently any theory dependent upon them is a futile exercise." Kimura responds that nonetheless "these quantities must be investigated and measured if the mechanisms of evolution are to be understood." Read carefully, Kimura's logic gives us no reason to suppose that the "mechanisms of evolution" actually can be understood by scientific investigation, since he has no real response to the criticism that it is impossible to establish the essential facts about such matters as population sizes and selection coefficients in the distant past. On the other hand, Kimura rightly points out that untestability is also a valid charge against selectionist theories, "which can invoke special kinds of selection to fit special circumstances and which usually fail to make quantitative predictions."

An example reported by Kimura illustrates the flavor of the neutralist-selectionist debate. The neutral theory predicted near 100 percent protein heterozygosity in a large population. Francisco Ayala reported that heterozygosity in a large-population fruitfly species was 18 percent, and so the neutral theory was wrong. No problem, responded Kimura: the discrepancy could be resolved by assuming that there was a population bottleneck of the right size sometime (maybe caused by the last ice age), or by adjusting the assumptions of the mathematical model in some other respect. Anyway, the selectionists were having their own problems explaining why natural selection would preserve as much heterozygosity as apparently exists. Both sides to the controversy assumed that either the neutralist or selectionist version of Darwinism must be true, and so each side could buttress its own case by disproving the other.

The articles quoted in footnote 2 are Roger Lewin, "Molecular Clocks Run Out of Time," *New Scientist*, 10 February 1990, p. 38; and Allan Wilson's previously cited *Scientific American* article.

Christian Schwabe expressed what I consider to be an appropriately skeptical view of molecular evolutionary theories in his article "On the Validity of Molecular Evolution" in *Trends in Biochemical Sciences*, 1986, vol. 11, pp. 280–82. He remarked that "it seems disconcerting that many exceptions exist to the orderly progression of species as determined by molecular homologies; so many in fact that the exception[s], the quirks, may carry the more important message." Schwabe complained of the frequent use of ad hoc hypotheses to reconcile discrepant molecular data with neo-Darwinism, and noted that "The neo-Darwinian hypothesis . . . allows one to interpret simple sequence differences such as to represent complex processes, namely gene duplications, mutations, deletions and insertions, without offering the slightest possibility of proof, either in practice or in principle."

One reason it may be unwise to draw conclusions about evolution from the molecular data is that molecular evolution is a relatively new field, and more detailed follow-up reports may call into question some of the results reported by enthusiastic pioneers. For example, the September 1989 issue of *Evolutionary Biology* contains an article by the German biochemist Siegfried Scherer, titled "The Protein Molecular Clock: Time for a Re-evaluation." Scherer studied ten different proteins representing more than 500 individual amino acid sequences. He reported that in no case were the data consistent with predictions based on the clock concept, and concluded that "the protein molecular clock hypothesis must be rejected."

Edey and Johanson's *Blueprints* does a good job at the popular level of explaining the archaebacteria, the molecular clock, and the impact of the molecular approach upon paleoanthropology. Of course, these authors do not question the Darwinist preconceptions.

Chapter Eight   Prebiological Evolution

For general background on prebiological evolution I particularly recommend the following books: A. G. Cairns-Smith, *Seven Clues to the Origin of Life* (1985); Robert Shapiro, *Origins: A Skeptic's Guide to the Creation of Life on Earth* (1986); and Charles Thaxton, Walter Bradley, and Roger Olsen, *The Mystery of Life's Origin* (1984). Cairn-Smiths and Shapiro are chemists with stature in the field. Both are gifted popularizers who candidly reveal that the problems of explaining the origin of life have often been underestimated as investigators have exaggerated the importance of minor successes. Both affirm the existence of a naturalistic solution as a matter of faith. *The Mystery of Life's Origin* was a pathbreaking skeptical account of the field that appeared while such as Carl Sagan were busy assuring the public that the problem was virtually solved. It has been given a cold shoulder by many because it explicitly considers the case for intelligent creation. It is very much up to the technical standard of the field, however, and may be too demanding for readers lacking a background in chemistry. Francis Crick's book *Life Itself* (1981) is inferior to the competition, despite the fame of its author, but the description of directed pan-spermia is not to be missed. For those who prefer a more earth-bound approach, the experimental and theoretical work of Manfred Eigen's group on the RNA "naked gene" is described in Edey and Johanson's *Blueprints*.

There is a good brief skeptical treatment of prebiological evolution in Chapter Eleven of Michael Denton's *Evolution: A Theory in Crisis* (1985). Carl Sagan's conclusion that the spontaneous origin of life must be highly probable because it happened in so brief a period on the early earth is quoted on p. 352 of Denton. Sagan's "start from the preferred conclusion and work backwards" logic is typical for workers in this field. For example, some scientists have refused to credit evidence that the early earth's atmosphere was not of the strongly reducing nature presupposed by the

Miller-Urey experiment, reasoning that the conditions necessary for the spontaneous production of amino acids must have been present because otherwise life would not exist. Robert Shapiro commented that "We have reached a situation where a theory has been accepted as a fact by some, and possible contrary evidence is shunted aside. This condition, of course, again describes mythology rather than science."

For an excellent brief overview of the field for the professional scientist I recommend the article "The Origin of Life: More Questions than Answers," by Klause Dose, in *Interdisciplinary Science Reviews*, vol. 13, no. 4, p. 348 (1988). See also, the brief review by Dose of a collection of papers about the mineral origin of life thesis, appearing in *Bio Systems*, vol. 22 (1), p. 89 (1988). Dose, a leading figure in prebiological evolution, is Director of the Institute for Biochemistry at the Johannes Gutenberg University in Mainz, Germany.

The article quoted in the text by Gerald F. Joyce, "RNA Evolution and the Origins of Life," appeared in *Nature*, vol. 338, pp. 217–24 (March 16, 1989). Joyce ended with the somber observation that origin of life researchers have grown accustomed to a "lack of relevant experimental data" and a high level of frustration.

Richard Dawkins' Chapter Six on "Origins and Miracles" in *The Blind Watchmaker* is a virtuoso piece of Darwinist advocacy, paying particular attention to Cairns-Smith's clay evolution scenario. Dawkins made use of Hoyle's "junkyard" metaphor to explain how a micromutation in the genes regulating embryonic development might produce additional ribs, muscles, and so on in the adult organism. The mutation would just be adding more of what already was in the program, and so Dawkins thought it would be a "stretch DC-8" mutation rather than a "Boeing 747" mutation. He considers it much more probable that a tornado hitting a standard DC-8 in a junkyard might transform it into a stretched version of the same airplane than that a tornado could convert pure junk into a 747.

The research involving computer models of self-organizing systems is most completely reported in two collections of papers reflecting conferences held in 1987 and 1990 at the Los Alamos National Laboratory. The 1990 conference is reported in the article "Spontaneous Order, Evolution, and Life," in *Science*, 30 March 1990, p. 1543. This is the article quoted in the text.

I have also benefitted from two unpublished papers by Charles Thaxton: "DNA, Design and the Origin of Life" (1986); and "In Pursuit of Intelligent Causes: Some Historical Background" (1988).

Chapter Nine   **The Rules of Science**

The legal citation to the opinion by Judge Overton is McLean v. Arkansas Board of Education, 1529 F.Supp. 1255 (W.D. Ark. 1982). The opinion is reprinted in the collection *But Is It Science?* (Ruse, ed., 1988). This collection also contains articles critical of the Ruse-Overton definition by the philosophers Larry Laudan and Philip Quinn, accompanied by replies from Ruse. For additional accounts of the trial by participants, see Langdon Gilkey's *Creationism on Trial: Evolution and God at Little Rock* (1985), and Robert V. Gentry's *Creation's Tiny Mystery* (2d ed. 1988). Gilkey is a liberal theologian who testified for the plaintiffs; Gentry is a physicist and a creation-scientist who testified in defense of the statute.

Stephen Jay Gould praised the opinion in the following terms: "Judge Overton's brilliant and beautifully crafted decision is the finest legal document ever written about this question—far surpassing anything that the Scopes trial generated, or any opinions [in the two other cases that went to the Supreme Court]. Judge Overton's definitions of science are so cogent and so clearly expressed that we can use his words as a model for our own proceedings. *Science*, the leading journal of American professional science, published Judge Overton's decision verbatim as a major article." ("Postscript," *Natural History*, November 1987, p. 26.)

Media accounts and judicial opinions take for granted that the balanced treatment statutes were the work of a highly organized nationwide coalition of creation-scientists, but this has been denied. According to the creation-scientist attorney Wendell R. Bird, most of the national creation-science organizations oppose legislation of this kind, "preferring instead to persuade teachers and administrators of the scientific merit of the theory of creation without legal compulsion." An individual named Paul Ellwanger appears to have taken the lead in proposing balanced treatment legislation, with the result that some reluctant creation-scientists were drawn into losing battles on ground not of their own choosing. See Wendell R. Bird, *The Origin of Species Revisited*, vol. 2, pp. 357–359 (1989).

The quotations from Thomas Kuhn's *The Structure of Scientific Revolutions* (2d ed. 1970), are from pages 5, 24, 77–79, and 127–128. Interestingly, Kuhn's model of the scientific enterprise is itself based upon Darwinist philosophy. Kuhn noted that the distinctive feature of Darwin's theory, from a philosophical point of view, was that it abolished the notion that evolution is a goal-directed process. Natural selection has no goal, but it nonetheless produces progress in the form of marvelously adapted organs like the eye and hand. Similarly, science progresses by "the selection by conflict within the scientific community of the fittest way to practice future science. The net result of a sequence of such revolutionary selections, separated by periods of normal research, is the wonderfully adapted set of instruments we call modern scientific knowledge. . . . And the entire process may have occurred, as we now suppose biological evolution did, without benefit of a set goal, a permanent fixed scientific truth, of which each stage in the development of scientific knowledge is a better exemplar." (pp. 172–173.)

The passage from Heinz Pagels' *The Dreams of Reason* (1988) is from pp. 156–58. The two quoted paragraphs are separated by three paragraphs in which Pagels discusses the logic of mathematics as an additional example of the cosmic building code of the Demiurge. The passages by George Gaylord Simpson are from *The Meaning of Evolution* (rev. ed. 1967), pp. 279, 344–45. Although Karl Popper's falsifiability criterion is unsatisfactory as a definition of "science," Popper's writing on this subject is extremely valuable for its insights into the difference between science and pseudoscience. This is the subject of Chapter Twelve.

Chapter Ten   **Darwinist Religion**

The 1984 statement of the National Academy of Sciences and Gould's reply to Irving Kristol are described in the research notes to Chapter One. Gould rebutted Kristol's charge that textbooks on evolution have an antireligious bias by citing the evident fairness of the authors of the leading textbooks, Dobzhansky and Futuyma. The naturalistic interpretation of "fairness towards religion" does not inhibit scientists from making explicit their assumption that theistic religion is nonsense. Here is what Futuyma has to say on pp. 12–13 of *Science on Trial: The Case for Evolution* (1983):

> Anyone who believes in Genesis as a literal description of history must hold a world view that is entirely incompatible with the idea of evolution, not to speak of science itself. . . . Where science insists on material, mechanistic causes that can be understood by physics and chemistry, the literal believer in Genesis invokes unknowable supernatural forces.
>
> Perhaps more importantly, if the world and its creatures developed purely by material, physical forces, it could not have been designed and has no purpose or goal. The fundamentalist, in contrast, believes that everything in the world, every species and every characteristic of every species, was designed by an intelligent, purposeful artificer, and that it was made for a purpose. Nowhere does this contrast apply with more force than to the human species. Some shrink from the conclusion that the human species was not designed, has no purpose, and is the product of mere mechanical mechanisms—but this seems to be the message of evolution.

William Provine's paper "Evolution and the Foundation of Ethics" appeared in *MBL Science* (a publication of the Marine Biological Laboratory at Woods Hole, Massachusetts), vol. 3, no. 1, pp. 25–29. A shorter version appeared as a guest editorial in the September 5, 1988, issue of *The Scientist*, with correspondence and rebuttals in succeeding issues. Provine also lectured on this theme at a major gathering of evolutionary biologists at the Field Museum in Chicago in 1987.

The booklet "Teaching Science in a Climate of Controversy" is available from the American Scientific Affiliation, P.O. Box 668, Ipswich, MA 01938-9980. The 1989 edition has been painstakingly revised to meet various objections, fair and unfair, to earlier versions. The Darwinist reviews quoted in the text appeared in the journal *The Science Teacher* for February and September 1987.

The quotation from Julian Huxley's *Religion Without Revelation* (1958) is from page 194. Many scientists have promoted ethical or inspirational philosophies based on evolution. For the depressing details see Mary Midgely's *Evolution as a Religion* (1986), and the essays in John C. Greene's collection *Science, Ideology and World View* (1981). I especially recommend Marjorie Grene's article "The Faith of Darwinism," in *Encounter*, vol. 74, pp. 48–56 (1959), whose theme is that "It is as a *religion of science* that Darwinism chiefly held, and holds, men's minds."

Dobzhansky's endorsement of Teilhard de Chardin's philosophy comes at the end of his 1962 book, *Mankind Evolving* (Bantam ed., 1970). The Teilhard quotes are from *The Phenomenon of Man* (1959). Dobzhansky described Teilhard's faith as "undemonstrable by scientifically established facts" but not contradicted by any scientific knowledge, and as a "ray of hope" for modern man which "fits the requirements of our times."

Teilhard de Chardin's aspiration to reformulate the Catholic faith with evolution at its center illustrates the difficulty of disentangling religious and scientific motives on both sides of the evolution controversy. Teilhard was not only a theologian but a major figure in paleoanthropology. He was closely involved with the amateur fossil hunter Charles Dawson and Sir Arthur Smith Woodward in the discovery of the fraudulent "Piltdown Man" in 1912–13.

There are strong grounds for suspecting that Teilhard's religious enthusiasm for evolution led him into participation in fraud. Many persons familiar with the evidence (including Stephen Jay Gould and Louis Leakey) have concluded that Teilhard was probably culpably involved in preparing the Piltdown fraud, although the evidence is not conclusive and

Teilhard's admirers insist that he was too saintly a man to consider such a thing. Gould's essays "The Piltdown Conspiracy" and "A Reply to Critics" in *Hen's Teeth and Horse's Toes* (1983) provide a good introduction to the subject. See also the Research Notes to Chapter Six.

Piltdown Man became an anomaly after the discovery of "Peking Man" in China in the 1930s (in which Teilhard also played an important role) led the experts to hypothesize a different path of evolution for early man, and retesting eventually established in 1953 that the skull skillfully combined the jaw of an orangutan with the skull of a modern man. Until the Piltdown fossil became inconvenient, after the British scientists who received the credit for its discovery had passed from the scene, the skull was guarded from skeptical investigators in a safe in the British Natural History Museum. Considering that some knowledgeable scientists had expressed skepticism about Piltdown Man from the time of its discovery, this concealment of the evidence is a greater scandal than the original fraud.

Chapter Eleven   **Darwinist Education**

The story of the controversy at the British Natural History Museum is mostly from the editorial and correspondence pages of *Nature* for 1980–1982, volumes 288–291. L. B. Halstead's letters appeared at vol. 288, p. 208; vol. 289, pp. 106, 742; and vol. 292, p. 403. *Nature*'s first editorial, "Darwin's Death in South Kensington," appeared in the issue of February 26, 1981, vol. 289, p. 735. The letter of response from the Museum's 22 scientists is in vol. 290, p. 82. The follow-up editorial "How True is the Theory of Evolution" is in vol. 290, p. 75. The final editorial word was delivered in a signed article by Barry Cox, vol. 291, p. 373. Gareth Nelson's letter is in vol. 289, p. 627.

Additional accounts of the Museum controversy can be found in Anthony Flew, *Darwinian Evolution*, pp. 33–34; Alan Hayward, *Creation and Evolution: Some Facts and Fallacies*, pp. 1–2 (1985); and Francis Hitching, *The Neck of the Giraffe*, pp. 219–23. The interview with the Museum's Director of Public Services, Dr. Roger Miles, is reported in Hitching, pp. 222–23.

The lecture by Michael Ruse titled "The Ideology of Darwinism" was delivered at a UNESCO-sponsored conference in East Germany in 1981, and published in English under the auspices of the Akademie der Wissenschaften der DDR in January 1983.

The *Science Framework* (for California public schools) was published by the California State Board of Education in 1990. The published version contains the *Policy Statement on the Teaching of Natural Sciences*, which was adopted by the Board in 1989 to supersede the Board's 1972 *Antidogmatism Policy*. The cytochrome c table appears in the *Framework* at page 116; the figures in this table were copied verbatim from *Of Pandas and People*, p. 37, by Percival Davis and Dean H. Kenyon, with Charles Thaxton (Haughton, 1989). This book is "creationist" only in the sense that it juxtaposes a paradigm of "intelligent design" with the dominant paradigm of (naturalistic) evolution, and makes the case for the former. It does not rely upon the authority of the Bible, and indeed its methodology is far more empirical than that of the *Framework*.

Chapter Twelve   **Science and Pseudoscience**

Popper's essay "Science: Conjectures and Refutations," from the collection *Conjectures and Refutations* (1963), is the principal source for this chapter. Bryan Magee's short book *Popper* (1973), provides a lucid summary of Popper's philosophy for the general reader. The quotation from Douglas Futuyma is from the opening chapter of his textbook *Evolutionary Biology* (1986). The Julian Huxley quotation is from volume 3 of *Evolution after Darwin*, (Tax ed., 1960), the record of the University of Chicago Centennial Celebration of the publication of *The Origin of Species*.

The text observes that Darwinism so fit the spirit of its age that the theory attracted a surprising amount of support from religious leaders. Many of Darwin's early supporters were either clergymen or devout laymen, including his most prominent American advocate, the Congregationalist Harvard Professor Asa Gray. Supporters of "evolution" included not just persons we would think of as religious liberals, but conservative Evangelicals such as Princeton Theological Seminary Professor Benjamin Warfield. Two specific factors influenced this support: (1) religious intellectuals were determined not to repeat the scandal of the Galileo persecution; and (2) with the aid of a little self-deception, Darwinism could be interpreted as "creation wholesale" by a progress-minded Deity acting through rationally accessible secondary causes. On the surprising receptivity of conservative theologians to Darwinism, see David N. Livingstone's *Darwin's Forgotten Defenders: The Encounter Between Evangelical Theology and Evolutionary Thought* (1987).

Index

Only substantive mentions included. (n) means footnote.